T0140407

International Series on Computer Entertainment and Media Technology

Series Editor

Newton Lee
Institute for Education, Research, and Scholarships
Los Angeles, CA, USA

The International Series on Computer Entertainment and Media Technology presents forward-looking ideas, cutting-edge research, and in-depth case studies across a wide spectrum of entertainment and media technology. The series covers a range of content from professional to academic. Entertainment Technology includes computer games, electronic toys, scenery fabrication, theatrical property, costume, lighting, sound, video, music, show control, animation, animatronics, interactive environments, computer simulation, visual effects, augmented reality, and virtual reality. Media Technology includes art media, print media, digital media, electronic media, big data, asset management, signal processing, data recording, data storage, data transmission, media psychology, wearable devices, robotics, and physical computing.

More information about this series at http://www.springer.com/series/13820

Farkhund Iqbal • Mourad Debbabi
Benjamin C. M. Fung

Machine Learning for Authorship Attribution and Cyber Forensics

 Springer

Farkhund Iqbal
Zayed University
Abu Dhabi, United Arab Emirates

Benjamin C. M. Fung
School of Information Studies
McGill University
Montreal, QC, Canada

Mourad Debbabi
School of Engineering & Computer Science
Concordia University
Montreal, QC, Canada

ISSN 2364-947X   ISSN 2364-9488 (electronic)
International Series on Computer Entertainment and Media Technology
ISBN 978-3-030-61677-9   ISBN 978-3-030-61675-5 (eBook)
https://doi.org/10.1007/978-3-030-61675-5

This Springer imprint is published by the registered company Springer Nature Switzerland AG
The registered company address is: Gewerbestrasse 11, 6330 Cham, Switzerland

Acknowledgments

The authors would like to sincerely thank Dr. Rachid Hadjidj and Dr. Hamad BinSalleeh for their contribution in the early stage of our collaborative projects. Some of the foundational concepts in our co-authored publications are further elaborated in this book. The collaborative work with Michael Schmid, Adam Szporer, Liaquat A. Khan, and Mirza Ahmed further strengthened the route towards idea of drafting this book. Special thanks go to Rabia Batool and Dr. Benjamin Yankson for their help in improving the structure and formatting of the content. Finally, the authors convey their gratitude to Dr. Patrick C. K. Hung, Laura Rafferty, and Dr. Aine Mac Dermott for their help in proofreading and editing the manuscript.

Contents

Chapter 1
Cybersecurity And Cybercrime Investigation

Society's increasing reliance on technology, fueled by a growing desire for increased connectivity (given the increased productivity, efficiency, and availability to name a few motivations) has helped give rise to the compounded growth of electronic data. The increasing adoption of various technologies has driven the need to protect said technologies as well as the massive amount of electronic data produced by them. Almost every type of new technology created today, from homes and cars to fridges, toys, and stoves, is designed as a smart device, generating data as an auxiliary function. These devices are all now part of the Internet of Things (IoT), which is comprised of devices that have embedded sensors, networking capabilities, and features that can generate significant amounts of data. Not only has society seen a dramatic rise in the use of IoT devices, but there has also been a marked evolution in the way that businesses use these technologies to deliver goods and services. These include banking, shopping, and procedure-driven processes. These enhanced approaches to delivering added value create avenues for misuse and increase the potential for criminal activities by utilizing the digital information generated for malicious purposes. This threat requires protecting this information from unauthorized access, as this data (ranging from sensitive personal data, demographic data, business data, to system data and context data) can be monetized by criminals.

There is an increased desire by the criminal enterprise to gain access to sensitive personal or business information, and/or the system that host it, in order to exploit it for financial (or other) gains. In this context, two challenges emerge. Firstly, there is the challenge of designing good cybersecurity programs to mitigate the intrusion risk posed by criminal enterprises. Secondly, in cases where security is breached, and a crime is committed (either for financial gain, cyberwarfare, or a crime in the form of online harassment, stalking, etc.), there must exist robust methods to forensically investigate such cybercrime and trace its origins back to the perpetrators.

© The Editor(s) (if applicable) and The Author(s), under exclusive license to
Springer Nature Switzerland AG 2020
F. Iqbal et al., *Machine Learning for Authorship Attribution and Cyber
Forensics*, International Series on Computer Entertainment and Media
Technology, https://doi.org/10.1007/978-3-030-61675-5_1

Cybercriminals abuse the Internet and associated web technologies by conducting illegal activities such as identity theft, fraud, critical infrastructure damage, harassment, intellectual property theft, hacking, bullying, etc. These activities often involve social engineering or electronic communication exchange among the criminals themselves. Successful forensic analysis of social media and online message posts for the purpose of collecting empirical evidence is necessary to identify the masterminds (e.g., rogue nations) and prosecute cybercriminals in a court of law and, when effective, this can be a way to limit future cybercrimes. One significant challenge in this task is the need to develop suitable tools and techniques in order to interpret and forensically analyze large volumes of suspicious online messages.

Digital forensics is a useful investigative branch of criminal law and involves analyzing, reporting, and presenting digital evidence and content, mathematical validation, the use of validation tools, and repeatability, to recreate a crime [1]. In an investigation of digital crime, it is important for an investigator to consider, the collection and analysis of electronically stored information (ESI) in a manner consistent with sound forensic principles. In doing so, efforts are made to prevent alteration of original evidence, the inclusion of complete collection documentation, preservation, and analysis of ESI, etc., while ensuring appropriate levels of restriction on who has access to the collected ESI [1]. Adopting and applying the best measures and forensic methodologies to solve cybercrime is an important part of the solution. In this book, different forensic analysis approaches are presented to help investigators analyze and interpret the textual content of online messages. The first objective involves the use of authorship analysis techniques to collect patterns of authorial attributes, in order to address the problem of anonymity in online communication. The second objective utilizes the application of knowledge discovery and semantic analysis techniques to identify criminal networks and illegal activities. The focus of such approaches is to collect credible, and interpretable evidence for both technical and non-technical professional experts (including law enforcement, intelligence organizations, personnel, and jury members).

1.1 Cybersecurity

Cybersecurity focuses on the body of technologies, processes, and practices intended to protect interconnected devices, programs, data, and network from cyberattack. It is as critical to business, government, and society today as regular security (i.e., policing) was 50 years ago. It is vital for organizations to recognize the importance of cybersecurity and to understand their responsibilities in securing personal and/or organizational infrastructure, which underpins broader digital connectivity. In a sense, cybersecurity is a framework to protect information and infrastructure, otherwise described as the vehicle that stores and transports electronically communicated information from one point to another. As day-to-day communications and practices become more electronically integrated and fluid, there is a growing need for members of society to better understand cybersecurity. This includes grasping a broad range of issues such as the protection of data, controls on infrastructure which store, transmit, and process information, as well as the ability to clearly identify and analyze cases where these controls have been breached and to correctly pinpoint the assailant. Cybersecurity is no longer the sole responsibility of a discrete group of trustworthy and reliable professionals; it is now the responsibility of everyone.

Societies have, over the years, developed overarching structures and frameworks to ensure that they function well. Frameworks (e.g., laws and rules) are usually applied to and generally followed by groups of people that can be easily identified within a geographically restricted area. This means that a country will have frameworks that govern its land and people, and these frameworks may differ from those of other countries. In essence, if any member of a given society engages in a malicious activity which contravenes the laws, regulations, and/or contracts of that society, the authorities are, in many cases, able to identify the perpetrator of the crime and gather the requisite evidence to prosecute them in a court of law. In the digital and cyber world, however, the concept of geographical boundaries does not exist. As a result, there is usually a challenge when some individual or organization commits a crime outside of the legal and regulatory jurisdiction of the affected party. This makes the expectation of abiding by laws or regulations (or enforcing them and prosecuting the perpetrator(s)) very challenging. Because of the complexity of law enforcement, application of policies, and identification of crime, a clear forensic process must be employed by investigators or authorities to effectively investigate online crime. Finally, based on the skill level of the criminal—in terms of covering their tracks—collecting enough evidence to analyze, identify, and effectively prosecute can range in difficulty from simple to nearly impossible.

The primary mission of a cybersecurity program is to ensure that cyber assets, information, and the systems that house them remain safe and useful [2]. As results of inherent flaws or weaknesses in system security procedures, design, and/or implementation, it is critical to make sure that any cyber asset which stores, transmits, or processes data is protected from existing and anticipated threats and vulnerabilities. Additionally, individuals and organizations must make concerted efforts to adhere to their legal and ethical responsibilities properly when understanding their role in the protection of systems and data. To do this well requires a desire to master the threat and vulnerability landscape, to understand related risks to their assets, and to address identified vulnerabilities, threats, and security risks. The protection of an organization's cyber-infrastructure relies heavily on managerial, technical, cyber-intelligence, and operational controls. Technical and operational controls employ advanced technology to provide safeguards on data or assets that are in process, in transit or stored. Managerial controls address the security issue of digital assets through administrative and managerial processes and procedures.

As part of the inherent rationale for cybersecurity, information has value. This value lies in the characteristics that the information possesses, and the degree to which it possesses given characteristics, in turn, defines its value. These characteristics are as follows [3]:

- **Confidentiality**—The prevention of unauthorized access to information by individuals or other systems.
- **Integrity** —The maintenance of the information's consistency, accuracy, and trustworthiness. The exposure of information to corruption, unauthorized change, destruction, damage, or other disruption of authenticity directly affects its integrity.
- **Availability**—The ability of authorized users, to access, as needed, information without interference, obstruction, or impediment.
- **Authenticity**—The degree to which information has maintained its genuineness or originality with respect to its creation (created, placed, stored, or transferred).
- **Accuracy**—The degree to which the information has the value that users expect and is free from unauthorized modifications, mistakes, or errors.

These characteristics are intrinsically important to the corporations, government agencies, or individuals that create or consume information as part of their activities. With such dependencies comes the obligation to understand the potential risks to their cyber assets, which include but are not limited to financial losses, critical infrastructure damage, reputational costs, legal and regulatory obligations, such as the Freedom of Information Protection of Privacy Act (FIPPA), General Data Protection Regulation (GDPR), Personal Health Information Protection Act (PHIPA), and International Organization for Standardization (ISO) 27,032, to name a few. Due to personal and societal reliance on technology in daily life, the likelihood of being the victim of a cyber-attack is much greater than it used to be. This is in part due to the ease with which sophisticated attacks can be launched via simple tools, the speed with which vulnerabilities in cyber assets are detected (and the lag before the vulnerabilities are patched), and concerns about ubiquitous malicious software and connected devices [3]. Rather surprisingly, an additional key factor is a low rate at which patches are downloaded and installed even when they are released.

1.2 Key Components to Minimizing Cybercrimes

There are specialized core areas that must be robust for a cybersecurity program to be effective in minimizing the risk of a successful cyber-attack. These include, but are not limited to [4]:

- **Hardware Security**: Hardware security is the protection of physical hardware from being stolen or surreptitiously replaced. Physical security and emanation security are two kinds of hardware security. Physical security relates to the protection of computer hardware and associated equipment, such as computers, servers, networks, connectivity, and peripheral equipment, etc. from security threats. These include tampering and theft, or natural disasters such as earthquakes, floods, etc. Emanation security encompasses constraints used to prevent signal emission by electrical equipment in the system, such as electromagnetic emission, visible emission, and audio emission.

- **Information Security:** Defense from vulnerabilities existing in the software or hardware architecture of the computer is called information security. Computer security and communication security are two types of information security. Computer security defines the working of access control mechanisms, hardware mechanisms that the operating system needs (e.g. virtual memory), and encryption mechanisms. It also defines how programs inside the computer should act to enforce security policies. Communication security relates to the security of data and information during communication.
- **Administration Security**: Administration security addresses the protection of information from vulnerabilities caused by users and threats against vulnerabilities in the security of the organization. Personnel security and operational security are two types of administration security. Personnel security encompasses procedures that maximize the chances that persons in the organization comply with security policies. In general, greater security risks are posed by authorized users (or internal users) than external attackers. Furthermore, an even larger threat can be posed by personnel responsible for maintaining security in a system if they are coerced to (e.g., via threats to members of their family living in another country) or willingly abuse their privileges. It is imperative to have protection mechanisms in place to deal with breaches by these types of users or at least have methods capable of limiting potential damage. On the other hand, operational security controls the implementation and operation of all other forms of security. It helps in the enforcement of the rules outlined in the security policy and outlines actions to take in case of security violations, the implementation of recovery mechanisms, etc.

1.3 Damage Resulting from Cybercrime

Over the past decade, there has been a tremendous increase in the use of internet-connected digital devices with the capacity to transmit, store, and process information; all stages of which need to be protected from unauthorized access/tampering. In 2012, at least 2.4 billion people had used Internet services. More than 60% of them were in developing countries, and 45% of them were below the age of 25 [6]. Fast forward to December 2017, and this number had skyrocketed to 4.2 billion people having access to the internet (while in 2020 it reached a figure of 4.8 billion[1]) [5]. Figures 1.1 and 1.2 show global internet use by geographic area and penetration rate respectfully. These figures are indicative of the non-proprietary and decentralized nature of the Internet, which offers ease of connectivity and provides access to a vast range of applications (i.e., Web, E-mail, File Storage Services, News Portals, VoIP, Gaming, Financial Services, e-Businesses and Social Networking). This unprecedented access has fundamentally changed daily life for most users.

---

[1] https://www.internetworldstats.com/stats.htm

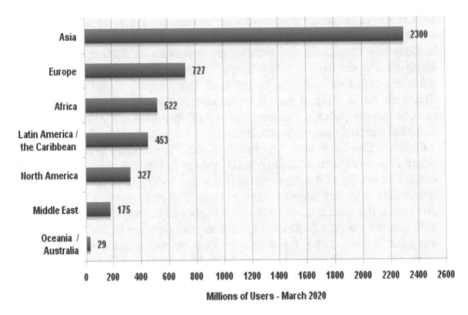

Fig. 1.1 Global internet use by geographic region—Mar. 2020 [5]

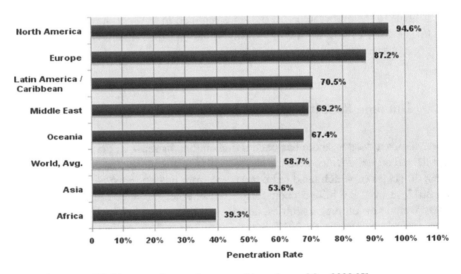

Fig. 1.2 Internet World penetration rate by geographic regions—Mar. 2020 [5]

Unsurprisingly, given the advancements in cellphone capabilities, mobile broadband subscriptions have seen impressive growth over the last few years, and continue to increase at present.

The 2016 projected estimate for the total number of mobile broadband subscriptions for 2017, with a value of 3.7 billion was surpassed by the end of 2017. According to a 2017 GSM study, 'Mobile Economy', the amount of unique mobile subscribers surpassed 5 billion globally [7]. Having exceeded, 5.2 billion people connected to mobile services in 2020,[2] the global mobile industry is expected to reach even greater milestones over the next few years. The number of unique mobile subscribers is projected to reach 5.9 billion by 2025, equivalent to 71% of the world's population [7]. By the end of the year 2020, we will live in a hyper-connected world of more than 80 billion interconnected IoT devices; this will truly be the Internet of the Future.

The Internet is the most attractive communications medium on the planet at present—and has also proven to be equally attractive for those who use it illicitly. In this era of hyper-connectedness, the Internet has unsurprisingly been employed by criminals and rogue nations to effectively perpetrate electronic crimes against ordinary people, various companies, and governments. As per the 12th annual Cost of Data Breach Study by IBM & Ponemon Institute [8] in 2017 the average cost per data breach was $3.62 million for participating companies. The average cost for a lost or stolen record containing sensitive information was $141.

These numbers are slightly lower than the 2016 report; however, the 2015 report showed that the total cost (on average) of a data breach for the participating companies was $3.79 million, and the cost per record was $217 to restore data [10]. A similar report conducted by PricewaterhouseCoopers (PWC) showed that organizations reported a 93% increase in financial losses as a result of breaches in 2014 as opposed to 2013. Importantly, although there was a decline in the overall cost in 2017 as compared to 2016 and 2015, companies in the 2017 study reported that companies were having larger breaches. The research shows a 1.8% increase in the average size of the data breaches, to more than 24,000 records [8].

Figure 1.3 shows the global average cost of cybercrime over the last 5 and respectively 6 years.

As illustrated, there was a steady increase in the cost of cybercrime from 2013 to 2015 and a significant increase in 2016, 2017, and 2018 with an average increase of 27.4% in 2016 [9]. Recently, the 2018 Global State of Information Security Survey [11] presented the responses of 9500 executives in 122 countries and contains valuable lessons from more than 75 industries worldwide. The respondents of the survey were keenly aware that the consequences of a successful cyber-attack include the disruption of operations, compromise of sensitive data, and damage to the quality of their products and services. The report further shows that cyber threats to the integrity of data are a rising concern, given their potential to undermine trusted systems (by manipulating or destroying data with or without the knowledge of owners) and can also damage critical infrastructures [12]. According to the report, 29% of respondents openly accepted loss or damage of internal records as a result of a security incident.

---

[2] https://www.gsma.com/mobileeconomy/wp-content/uploads/2020/03/GSMA_MobileEconomy2020_Global.pdf

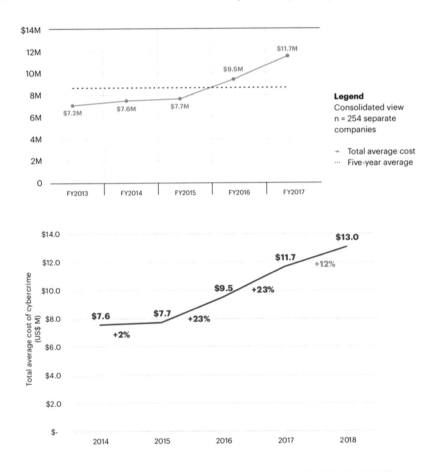

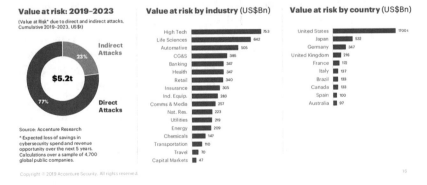

Fig. 1.3 The global average cost of cybercrime over 5 years [9]

For example, in 2016, 32% of customer records were compromised compared to 35% in 2017. Similarly, 30% of employee records were compromised in 2016 as compared to 32% in 2017 [12]. Another report from Dell [13] sheds more light on the companies' security breaches. The Dell study, entitled "Protecting the organization against the unknown—A new generation of threats", strengthens the concern with regards to corporate cybersecurity risks. Survey results indicated that nearly 3/4 of organizations reported a security breach within the last 12 months. Moreover, the economic value that is at risk due to cyber-security over the 5 year span 2019–2023 is $5.2 trillion dollars.[3]

Failing to successfully implement a robust security system with contingencies for different scenarios introduces tremendous risks to a business's long-term viability (along with expensive insurance policies) and can even make it impossible for companies to recover after an attack. To highlight some of the dangers of failing to take cybersecurity seriously, there are many documented cases of companies who never recovered from cyber-attacks. The following examples detail some medium-size companies that never recovered, illustrating the inherent dangers of cyber-attacks [14]:

- **Code Spaces:** Code Spaces, a former Software as a Service (SaaS) provider, was hacked and closed within 6 months after an attack [16].
- **MyBizHomepage:** The online company, valued at $100 million, faced a revenge attack after the chief technology officer and two other senior officers were fired; it cost the company over $1 million to resolve the breach [17].
- **DigiNotar:** the Dutch certificate authority (CA) declared bankruptcy after being at the center of a significant hacking case [14].

Figure 1.4, below, shows companies that have been hacked in relation to the amount of data loss suffered [18].

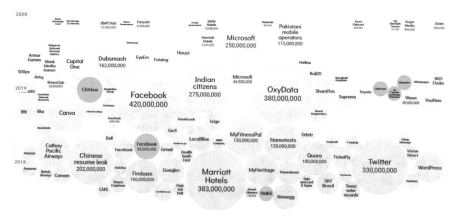

Fig. 1.4 World biggest data breaches, updated on Mar. 2020 [15]. https://www.informationisbeautiful.net/visualizations/worlds-biggest-data-breaches-hacks/

---

[3] https://www.slideshare.net/accenture/cost-of-cybercrime-study-in-financial-services-2019-report

The ideal way to mitigate cyber-attacks would be to use impenetrable security systems (e.g., data diodes) that could authenticate any data sent or accessed over the Internet. However, no security systems to date have proven effective in providing foolproof security. Perfect (100%) security is impossible but cybersecurity attempts to reduce the risk of attack and exploitation. In addition, the combination of general ignorance regarding potential threats as well as the enormous number of vulnerable software applications available on the Internet makes it more challenging to create a truly impenetrable cybersecurity solution. A survey sponsored by the Hewlett-Packard Company and conducted by the Ponemon Institute—"The 2013 Cyber Crime Study"—revealed that the rate of cybercrime is on the rise [19]. This study showed that enterprises from the sample pool spent, on average, approximately $11.56 million annually on cybersecurity. This figure is 26% larger than the previous year. Comparatively, a similar study in 2016 reported metrics on the number of attacks. This Global study [5] of 237 companies reported a total of 465 total attacks, resulting in an annualized average cost of approximately 9.5 million per incident. Additionally, a 21% net increase in the total cost was reported from 2015 to 2016. Sadly, these organizations are not new to cybersecurity; they spend large sums of money to buy commercial off-the-shelf security solutions for the detection and prevention of online crimes.

1.4 Cybercrimes

Rises in the integration of digital spaces and general connectedness have provided modern-day criminals and rogue nations with easy avenues to exploit online anonymity. Masquerading as others and using various web-based communication tools, cybercriminals can conduct illicit activities such as spamming, phishing, fraud, bullying, drug trafficking, and prostitution, all of which are difficult to trace back to the perpetrator of the crime. The increased use of networked computer systems and the sophistication of small portable devices has changed the world and simultaneously fueled the evolution of the criminal landscape. It has provided fringe members of society with an intricate and largely unregulated environment in which they can take advantage of the absence of comprehensive cybersecurity solutions, to cause harm to organizations and individuals. By common definition "cybercrimes" are occurrences of electronic acts that pose serious threat to the global economy, the safety of to the global economy, national critical infrastructures, and to the well-being of society as a whole [20].

Although laymen definitions exist there are no official, globally-consistent (or agreed upon) definitions of "cybercrime" available, due to complexities in the depth and scope of what can constitute a cybercrime [21]. Furthermore, there is no definition of cybercrime that appropriately distinguishes it from other forms of crime. However, there is a widespread and interchangeable series of words used to describe specific types of cybercrime, such as cyber warfare and cyber terrorism [22]. A collaborative attempt was made to define cybercrime at the Convention on

Cybercrime in 2001 [23]. As per the convention treaty, cybercrimes are defined as "crimes committed via the Internet and other computer networks, dealing particularly with infringements of copyright, computer-related fraud, child pornography, and violations of network security" [23].

Various legal jurisdictions are consistently having to adapt to technological innovations that provide exploitable opportunities for criminals. Cybercrime is a complex multi-jurisdictional problem, as neither cybercriminals nor their victims are restricted to a geographical region, social status, race, or economic status. It is well documented in national crime statistics reports and in surveys by law enforcement agencies and think-tanks that documented cases of cybercrimes are steadily on the rise [24]. For example, in the United States, as per the US Federal Bureau of Investigation (FBI), the Internet Crime Complaint Center (IC3) received 298,728 complaints of Internet crimes in 2016 [21]. This was made possible because the IC3 had set up a trustworthy avenue for reporting information concerning suspected cyber-criminal activities. From the complaints filed in 2016, there were approximately $1.33 billion in losses [25]. To provide context, since the inception of the FBI Internet Crime report in May 2000, it took 7 years before the IC3 received its one-millionth complaint. It only took an additional 3 years before the two million mark was passed in November 2010, while in 2018 close to 3.5 million complaints had been received. The complaints cite many different internet scams affecting the victims from around the globe [25], and as per the 2018 IC3 report, $2.7 billion in losses had been reported. These losses, while sizeable, are generally restricted only to reported statistics and do not include unreported incidents. Considering that the average per-complaint loss is $8421, it is not farfetched to assert that many cybercrimes resulting in individual losses below a threshold go unreported. It is important to note that many cybercrimes like bullying, sex-trafficking, drug trafficking, etc., do not have any associated quantifiable financial value.

Although there are varying degrees of behaviors or actions which can be classified as online crimes, there are specific types of cybercrimes that are frequently reported. Without a common understanding of what these types are, neither victims nor perpetrators can make truly informed, intelligent, and/or moral decisions in handling or policing cybercriminal activities worldwide.

1.4.1 Major Categories of Cybercrime

The following provides high-level information on various forms of cybercrime.[4] Some of the following are more elaborately described in the 2017 IC3 annual report [21]:

---

[4] Sarah Gordon, "On the definition and classification of cybercrime", https://link.springer.com/article/10.1007/s11416-006-0015-z

- **Identity Theft and Financial fraud:**
  - **Identity Theft**: Identity theft occurs when personal information, such as one's name or social security number, is used to impersonate an unsuspecting individual for the purposes of committing fraud by gaining access to accounts (Account Takeover). Credit card fraud is identity theft in its simplest form. It is a broad term for theft and fraud committed using a credit card, debit card, or any similar payment mechanism such as automated clearing house (ACH), electronic funds transfer, recurring charge, etc. The limited restriction and verification processes present during many online transactions make it easy for cybercriminals to conduct online credit card fraud. For example, recently a Mississippi woman was charged for illegally buying $140,000 in goods and services using the personal information of a 77 year old woman she assisted with bill payment and shopping [26].
  - **Charity fraud**: In charity fraud, perpetrators use deception to receive money from individuals who believe that they are making donations to legitimate charities helping the victims of disasters. This type of cybercrime is usually most prevalent during disasters or other instances of misfortune affecting a particular population. For instance, after Hurricane Harvey and Hurricane Irma in 2017, the National Center for Disaster Fraud received more than 400 complaints of fraud [27].
  - **419/Overpayment (419):** Solicitations from persons fraudulently seeking help in enabling the transfer of a considerable amount of money. The "sender" offers a share or commission in handsome funds subject to pay some nominal amount as cost or authorization fee etc. associated with the transfer. This type of crime has been well documented and associated with Nigerian 419, where there is intrinsic interest by mostly Nigerian and West African nationals to defraud westerners by offering them this type of deal in exchange for huge sums of money, gold, or inheritance. The "419" refers to the section of Nigeria's Criminal Code dealing with fraud. In instances of Overpayment, a renter or a buyer sends the targeted individual a cashier's check in an amount greater than that which was agreed to and then requests that the difference be wired back. Another variation of the Nigeria 419 is called the "advanced fee," where the targeted individual pays money to the fraudster in advance and expects to receive something of greater value in return but receives either nothing or receives significantly less than expected. In most of these cases, there is an activity trail which can include e-mail exchanges, money transfers, and possibly phone calls.

- **Cyber-terrorism:** Acts of violence that are intended to create a climate of terror or fear for religious, political, or ideological reasons and deliberately target or disregards the safety of non-combatants.
- **Cyber-extortion:** The practice of unlawfully obtaining money, services, or property from an individual through intimidation or abuses of authority. It can involve various threats including but not limited to physical harm, property damage, criminal prosecution, or public exposure.

- **Cyber-warfare:** Cyberwarfare represents the use of computer aided technologies to remotely carry out attacks on national critical infrastructure (e.g., utilities) and/or essential government and business services (e.g., healthcare, banking) in order to inflict harm comparable to actual warfare.

- **Computer as a target:**

  - **Corporate Data Breach:** This activity involves the unauthorized transfer of corporate data sent from a secure location to an untrusted environment (i.e. either an off-site computer or released to the internet) by accessing a computer or network, or by bypassing network security remotely. When a corporate data breach occurs, sensitive, protected, or confidential data is copied, transmitted, viewed, stolen, or used by an unauthorized individual or individuals. As demonstrated by Fig. 1.4, this type of cybercrime accounts for many of the world's largest data breaches.

  - **Denial of Service (DoS) and Distributed Denial of Service (DDoS) attacks:** In essence, DoS/DDoS involve the interruption of a legitimate system or network access to authorized users. These attacks are typically executed with malicious intent but occasionally do occur accidentally (for example, when one system sends too many consecutive requests to another, overloading the receiving system's capacity to effectively process further requests). This type of crime impacts the availability of organizational systems to legitimate users, resulting in significant decreases in the level of service provided by the organization to said users. With the IoT, this can also have serious physical implications to users in the case of Industrial Control Systems (ICS) or health monitoring systems.

  - **Business E-mail Compromise:** A scam targeting businesses that perform wire transfer payments and work with foreign suppliers. The scammers use social engineering or computer intrusion methods for the purposes of compromising business e-mail accounts to perform unauthorized transfers of funds. The degree of sophistication involved in executing crimes of this nature necessitates the involvement of a skilled hacker whose job is to understand the target company's infrastructure in enough detail to penetrate it. As a result of Business e-mail compromise, the Xoom corporation, a money exchange company transferred $30.8 million in corporate cash to fake foreign accounts.

  - **Malware/Scareware/Ransomware:** Malicious software designed to damage or disable the computers and/or computer systems of targeted victims. Software of this nature can be used for a variety of reasons, from obtaining private governmental information to gathering personal information such as identification numbers, credit card information, passwords, etc. Ransomware, specifically, is a type of malicious software that blocks a user's access to their computer system (mainly via encryption of data) and demands a sum of money be sent to the perpetrator.

- **Computer as a tool:**

  - **Cyberstalking:** The use of electronic communications or the internet to harass an individual, group, or organization. This can include false accusations, monitoring, threats, and other means of harassment. Cyberstalkers make use of social media, internet databases, and other online resources to follow and terrorize their victims.
  - **Virus dissemination:** The release of code that has the potential to propagate itself and damage, corrupt, or destroy digital systems.
  - **Phishing:** The act of sending a fraudulent e-mail that is made to look as if it comes from a trusted sender. Attackers try to gain personal information from the victim for financial gain, e.g., credit card information or financial status. Cybercrime attacks such as Advanced Persistent Threats (APTs) and ransomware often start with phishing [28].
  - **Software piracy:** Unauthorized use and distribution of computer software is called software piracy. This include, but not limited to, counterfeiting software, renting, shoplifting, OEM unbundling, Internet software piracy, hard disk loading, and corporate software piracy.
  - **Intellectual Property Rights (IPR)/copyright and counterfeit:** The unlawful appropriation of another's inventions, novel ideas, and/or creative expressions (Intellectual Property) is called intellectual property theft. Intellectual property can include trade secrets, proprietary product(s), or aspects of creative works such as movies, music, or software.

- **Obscene or offensive content:** Activities mostly related to the abuse and exploitation of children. Most of these types of crimes involve the distribution of different forms of visual content depicting children involved in acts (often sexual in nature) which are contrary to what is considered legally and/or societally acceptable. Such activities and distribution of related content occurs within chat rooms, peer-to-peer sharing networks, through e-mails, or dropsite hubs.
- **Drug trafficking:** Drug traffickers can use information technology to exchange encrypted messages in order to secretly carry out their illegal activities.

A global study published by PricewaterhouseCoopers shows that the number of cyber-related security incidents reported globally has increased by 48% during 2013; which is approximately an equivalent of 117,339 attacks per day [29]. The report not only highlights the increase in crime but also makes the case that cybersecurity breaches are becoming increasingly complex and impactful and thus are now capable of causing more destruction to organizations or individual victims than ever before. Cybercrime poses a serious threat to the global economy, critical infrastructure, and to the safety and well-being of society in general. It poses a serious threat to the operations of business and government organizations, and national economies and the global financial damage has been estimated at $225 billion [30].

Cybercrime can have a wide range of impacts on victims, which include but are not limited to, death or self-harm, cutting of access to critical infrastructure services (e.g., utilities, banking), identity theft, sexual exploitation (of children or otherwise), cyber-harassment, cyberbullying, intimidation, anxiety or other mental health illnesses, financial loss, and more. One tragic example of how severe the effects of cybercrime can be is that of 15 year old Amanda Todd, who in 2012 committed suicide after being severely bullied when someone posted nude images of her online [31].

Cybercrime attracts highly talented people, which can be recruited or coerced by cybercrime syndicates as well as entrenched fifth columns and rogue nations with sophisticated goals to target vulnerable systems, critical infrastructure, individual victims, governments and corporations. As such, the experts, researchers, and law enforcement or intelligence agents in this field must consistently enhance their understanding of criminal methodologies and capabilities to successfully track and thwart attempted attacks, and when cybercrimes occur, to gather enough evidence to facilitate the identification of the masterminds and prosecution of the perpetrators.

1.4.2   Causes of and Motivations for Cybercrime

There are many factors that can motivate an individual or group to commit cyber-crimes, several of which are described below.

1. **Personally motivated cybercrimes**
   Many cybercrimes are motivated by the emotions and desires of an individual. There can be many factors that influence these types of actions, not all of which are for financial gain. Some examples include teenagers who try to hack into their school's website, disgruntled employees who install malicious software on their employer's servers, teenagers who hack into their current or former inti-mate partners (e.g., boyfriend/girlfriend) social media accounts, etc. While some attacks of this nature are intended to be exploratory and not destructive, the potential for causing damage can be very high.
2. **Economically motivated cybercrimes**
   The opportunity for financial gain is another common motivation for cybercrimi-nals. There is a wide variety of approaches that a cybercriminal can take in mak-ing money from cybercrime. For example, when individuals are the targets, cybercriminals make use of different techniques such as phishing, identity theft, malware distribution, etc., to gain access to financial login information for the purposes of stealing money. To provide a sense of scope, according to [32], 25.7% of all malware attacks in 2018 targeted financial services and financial organisations. The actual figure is most likely much higher, as not all incidents are reported—or noticed. While large transfers are likely to be flagged by account owners as suspicious, some cybercriminals set up recurring monthly micro-transactions disguised as small purchases that are likely to go unnoticed. If the account holder does not inspect their transaction history regularly these charges can go unnoticed for years, and the total amount lost can grow significantly.

3. **Ideologically motivated cybercrimes**

Ideological beliefs held by an individual or a group can fuel some of the most dangerous cybercrimes. The motivation behind this type of cybercrimes originates from an ideological or geostrategic standpoint, and therefore the perpetrators can fervently believe or conclude that they are 'doing the right/morally just thing' on behalf of a given cause or to advance their agendas. Ideology-based cyber-attacks can be extremely destructive when they are fueled by a desire to cause the most damage possible to a given target or population.

4. **Access to confidential information**

Computer systems and online networks are utilized by financial organizations, security firms, and governmental organizations to store their sensitive and confidential information. The digital format of confidential information creates a potential attack vector for cybercriminals to exploit. Experienced cybercriminals can evade and bypass the controls implemented by state-of-the-art security systems or use social engineering and coercion of vulnerable insiders to obtain unauthorized access to confidential information for a variety of purposes.

5. **Types of Opportunistic Cybercrime**

   5.1 **Negligence/Carelessness**

   Cybercriminals exploit lapses in security resulting from user negligence or carelessness. Leaving sessions active or storing cookies and passwords on public networks, storing unsecured sensitive information, and leaving Wi-Fi access open without a password constitute a few examples where human negligence can facilitate the activities of malicious actors on the Internet.

   5.2 **Vulnerable Systems**

   Vulnerabilities in software systems can allow cybercriminals to gain unauthorized access to a vulnerable system. Common vulnerabilities in the most popular applications are usually released as Common Vulnerabilities and Exposures (CVE). In response, software vendors provide security patches in version updates or service packs. The time-lapse from the moment when a vulnerability is exposed, to the time when a fully functional patch is released can be used by the attackers to exploit the vulnerability on as many systems as possible. One of the most infamous cases to date of a vulnerability being exploited was the release of the "WannaCry" ransomware that infected many computers around the world in 2017. It spread rapidly across many computer networks—mostly those running the Windows Operating System—encrypting files, which made it impossible for authorized users to access the encrypted affected files [15]. The perpetrators demanded a ransom for the files to be decrypted.

6. **Lack of Evidence**

Evidence is required to convict an accused perpetrator in a court of law. If substantial, pertinent, and credible evidence cannot be brought against an accused perpetrator they are likely to walk free. In the case of cybercrimes, electronic evidence is required. Digital forensics helps in the generation of electronic evidence by identifying, collecting, and analyzing digital assets. This poses a great challenge for cybercrime investigators when seeking to properly authenticate and verify sources of information pertinent to electronic crime, as digital assets/ artifacts can easily be destroyed or altered. Experienced cybercriminals employ a variety of methods to hide, fabricate or destroy the trail of evidence detailing their illegal activities online, obfuscating investigators and police.

1.5   Major Challenges

The more we rely on the Internet, the more vulnerable we are to electronic crimes. Globally, a significant number of people have now become victims of cybercrimes such as hacking, interruption of essential services provided by critical infrastructures, identity theft, phishing, cyberstalking, online harassment, malicious software, etc. According to a report [33], nearly 73% of Americans have been affected by, or have attempted to engage in, some form of cybercrime. The increased incidence of cybercrime has forced legislators to rethink the enforcement of laws pertaining to Internet usage. The task is daunting, as there are major challenges with designing virtual infrastructure and processes which interface with and complement those governing society in the physical world, especially considering the cyber world's lack of geographic boundaries. As more and more societal functions migrate to the cyberworld, the occurrence of cybercrimes will likely increase exponentially, due to the ease with which they can be executed, their effectiveness, and the low probability that the perpetrator will face appropriate repercussions. As such, the challenge is now to figure out how to design efficient controls to protect citizens of the cyber world, and how to effectively investigate cybercrimes when they occur.

To secure the cyberworld, existing technologies, processes, and best practices must be used to protect and enhance the infrastructure and software that support the functionality of cyberspace. In the event of a cybercrime, it is imperative to identify the culprit in cases where the controls fail, and an environment is breached. To address some of the core issues in organizational cybersecurity, a well-rounded cybersecurity program and mitigation strategies for dealing with cybersecurity incidents are required. The goal of a cybersecurity program is, to provide the framework for a desired level of security by analyzing potential and realistic risks, putting in place best practices to mitigate them, and keeping the security practices updated. Other major challenges, discussed below, include developing cyber-security policies, finding technical and administrative controls to enforce these policies, implementing effective cybersecurity education and training programs, and having the ability to conduct pre-breach audits and/or to create post-breach incident reports which can help to identify criminals and/or exploited vulnerabilities. In the following section, major challenges influencing these topics are discussed.

1.5.1 Hacker Tools and Exploit Kits

Protecting oneself, let alone an organization, from cybercrime has become increasingly difficult due to continual advancements to, and the ever-evolving sophistication of, hacker tools and exploit kits. Many advanced tools and underground services available online make it easy to perpetrate cybercrimes. Exploit kits like Blackhole [31] or RIG [34] can be used by cybercriminals to compromise systems by exploiting the vulnerabilities found in applications installed on the target's machine. These tools and kits are sold and even rented on the black market and empower the attacker to easily launch an attack on the target's machine. Due to their ease of use and effectiveness, these tools are frequently employed by cybercriminals.

1.5.2 Universal Access

The world of the Internet has provided several convenient ways to connect people from diverse parts of the world on a single platform. This is especially true when we consider the advent of social networking, VoIP, e-business, financial services, etc. With a few mouse clicks or screen taps, virtually anyone can connect to other computers (and their operators), irrespective of geographic location. This has been made possible due to the decentralized nature of the Internet. Any machine or device connected to the Internet has the potential to be contacted and targeted. Usually, authentication systems and firewalls are employed to restrict unauthorized access to systems, but in many cases no access restriction is present, or the intruder can circumvent said restrictions. Similarly, unless encrypted, the information transmitted across the Internet can be seen by anyone with access to the network on which it traverses. While those authorized on the network may be able to view specific information, attackers can potentially view, edit, or even disrupt these network communications entirely. Several types of crimes such as eavesdropping on private communications, impersonation, IP or Web spoofing, intercepting or redirecting information to other destinations, denial of service attacks, etc. can be perpetrated by exploiting the open nature of the Internet.

1.5.3 Online Anonymity

Online documents or electronic discourses are written communications exchanged between people over the Internet. The mode of communication of online documents can be synchronous, such as chat logs, or asynchronous, such as e-mail messages and web forums [176]. During a synchronized communication, multiple parties are engaged at the same time and usually wait for replies from each other prior to response. An example of such an instance will be engaging a company with an online real time support chat session. In such instance, the customer support person will provide information about the product and the challenges the user is facing. The user will then wait for a response from the customer's service representative. Another avenue for synchronized communication is during a one-to-one or group chat. In synchronized communication, before messages are exchanged, and a synchronous scenario can begin, the sender and receiver need to establish a communications session and agree which party is going to be in control [177]. Once the session is established, the back and forth conversation occurs in real time, and usually participants await meaningful feedback before responding. When a chat message is sent, the party at the other end is waiting to receive and respond to the message. Both parties are working together at the same time with their clocks synchronized (time zone differences notwithstanding) [177]. On the other hand, asynchronous communication occurs when there is no avenue to activate real time chat, where the user sends a message and waits. In most instances, the corresponding party is neither expecting nor waiting for an incoming message and therefore may not view it right away. The two parties are not working together in real time and may even be unaware of the other's intended actions [177]. With both types of online communication, there may exist some level of interest in the identification of the authors of the messages exchanged.

Cybercriminals make use of different channels such as anonymous proxies, Internet cafes, and open wireless Internet access points to maintain their anonymity so that their real identities can remain hidden. As the first step in a cybercrime investigation, the IP address of the suspected perpetrator is usually identified. However, experienced criminals do not leave any trace of their own identities to be tracked down. The perpetrator will often take control of another victim's computing resources in order to commit a crime, making it difficult to attribute to the original point of attack. Similarly, they can also destroy all mediums used—e.g., disposing of hard disks and digital media, cleaning up logs and access histories, and using encryption and obfuscation for their online communication. These acts can remove all digital trails linking them to the crime committed.

1.5.4 Organized Crime

The globalization of the internet makes it a potential billion-dollar industry for illicit activities. The Internet, and by obvious extension the underlying hardware (e.g., computers, mobile devices), have now become the preferred medium for the perpetration of many serious crimes. The increased number of cybercrimes can be in-part attributed to the low risk of repercussions as well as to the high potential for rewards (financial or otherwise).

Unlike conventional criminal groups, cybercriminals act mostly on an individual basis, seldom coming into direct physical contact with one another. Most prefer to hide their identity and to only meet their group members online. The core members of the group usually plan and distribute the tasks and activities among the members of the group. Usually, individuals are assigned tasks based on who has the required skills to execute them, e.g., phishing, spamming, web defacement, identity theft, and decrypting codes.

The high level of anonymity and the decentralized nature of the Internet presents a significant challenge for law enforcement agencies to trace and locate cybercriminal groups. The group's activities can be organized and coordinated across a myriad of different geographic locations and they can maintain and mirror their servers in numerous different countries and jurisdictions, some of which may tolerate their presence since this can provide plausible deniability for state sponsored cyberwarfare activities). In order to cope with these types of intelligently organized cybercriminal groups a high level of international cooperation and coordination is often required, where the legislative authorities of several countries have to collectively synchronize efforts to bring the perpetrators to justice.

1.5.5 Nation State Threat Actors

Nation state threat actors are cybercriminals who are focused on breaching other nation's critical infrastructures, election processes, military, or diplomatic systems to access or steal extremely sensitive information for a competitive advantage over the target country [35]. These threat actors are not only interested in governmental data, but also use sophisticated techniques to disrupt private business or enterprise in order to influence executives business decisions, steal sensitive business documents or intellectual property, benefitting local or foreign competitors. According to the US National Intelligence's Worldwide Threat Assessment report released in 2018, Nation state threat actors are on the rise [36]. The report outlined grave concerns reported by Nation state and noted the high probability that incidences of this form of cybercrime will likely continue to increase. Documented incidents of nation state threat actors having significant impacts include, but are not limited to, Nation state-sponsored cyber-attacks in 2016 and 2017 against Ukraine and Saudi Arabia, both of which affected critical infrastructure, government, and commercial networks [36]. Also, due to the commercial availability of malware, a significant and alarming increase in ransomware and malware attacks have been seen globally [37].

1.6 Cybercrime Investigation

Tremendous Internet-related technological advancements have been made over the past 30 years, many of which present immense potential for the business community. The daunting task of effectively identifying and tracking online crime has made the investigation of cybercrimes difficult. Advanced technologies greatly help cybercriminals in obscuring their online activities, often leaving behind little or no digital footprints for investigators to follow. Furthermore, the tools, techniques, and technologies available to cybercriminals allow them to speed up the execution of online criminal acts, while often outpacing attempts by law enforcement agencies to track their activities. As such, cybercrime investigators should be equipped with both conventional investigative skills as well as a good understanding of the latest analysis tools available; in particular, those pertaining to advanced digital forensics.

Typically, a cybercrime investigation begins when the authorized investigation unit receives a victim's complaint. As the first course of action, a strategy and method to locate the source of the attack is chosen. Typically, this involves back-tracking through the chain of IP addresses in the hopes of identifying the source IP of the device (e.g., computer, mobile device) where from the attack originated. Once identified, authorities can sometimes (depending on the jurisdiction) obtain information specific to the activities of that IP address from the user activity logs. Such data is provided by the Internet Service Provider (ISP) used by the cybercriminal, in order to connect to the internet during the attack.

Identifying the perpetrator by analyzing the digital trails of information and providing robust and adequate evidence for the prosecution is a difficult job. Additionally, the urgency required during all investigative phases makes this task even more challenging. For example, very few ISPs keep records of their users for more than 30 days, and small ISPs may even delete them after 10 min. Retention of records is dependent on legal regulations (if they exist) for the particular jurisdiction. Some countries do require that ISPs maintain user logs for longer periods of time and that they provide other information to support ongoing investigations if requested with an appropriate cause.

To summarize, the surge in Internet popularity and growth of Internet-enabled devices has resulted in a simultaneous rise in cybercriminal activities, many of which can be particularly damaging to the individual, business, nation state or global economy. Digital processes aimed at investigating and preventing cybercrime are presently struggling to adapt to the rapidly evolving digital world of IoT. Said difficulties are compounded by the lack of global consensus on cybercriminal activity, with varying definitions and variable laws against cybercrime. As such, new procedures must be developed, and tools updated, in order to redress this imbalance, and prepare us for a more secure future of the digital world. Subsequent chapters explore these themes further, drawing attention towards these (and other) issues currently facing the domain of cybercrime and digital forensics.

Chapter 2
Messaging Forensics In Perspective

This chapter presents the central theme and a big picture of the methods and technologies covered in this book (see Fig. 2.2). For the readers to comprehend presented security and forensics issues, and associated solutions, the content is organized as components of a forensics analysis framework. The framework is employed to analyze online messages by integrating machine learning algorithms, natural language processing techniques, and social networking analysis techniques in order to help cybercrime investigation.

Cybercriminals frequently abuse the anonymity of online communication in order to conduct dubious and malicious online activities. These include phishing, spamming, identity theft, etc., which can often lead to further crimes such as harassment, fraud, cyberbullying, child pornography, and other region-specific crimes. These crimes can infringe upon the rights of citizens and can have other serious consequences. In phishing fraud, for instance, scammers send out messages and create websites to convince account holders to disclose their sensitive account information, such as account numbers and passwords. According to Wombat Security Technology, who compiles an annual "State of the Phish" report based on the data collected from millions of simulated phishing attacks, phishing remains a highly effective attack vector. Given that, it is unsurprising that phishing is responsible for a significant number of data breaches today [38]. As per the Phishing Activity Trend report, published by the Anti-Phishing working group (APWG) in February 2017, the total number of phishing attacks reported in 2016 was 1,220,523; 65% higher than reported in 2015.

In 2015, 85% of polled companies report being the victim of a phishing attack; a 13% increase from 2014. Additionally, 67% of the same companies polled reported spear phishing attacks; a 22% increase compared to 2014 [39]. The APWG Phishing Activity Trends Report collects and analyzes phishing attacks reported by its member companies and its Global Research Partners [39]. According to the report, phishing attack campaigns in 2016 shatters all

Fig. 2.1 Month-by-month comparison between 2015 and 2016 [38]

previous year's records. Figure 2.1 shows APWG comparative month by month records between 2015 and 2016.

The report also demonstrated that phishing attack templates that have the highest click rates incorporated items that employees expected to see in their work e-mails, such as Human Resource (HR) documents or shipping confirmations. The survey further demonstrates that employees were very cautious when they received e-mails from a consumer about non-work topics e.g., gift card notifications, social networking accounts, etc. However, "urgent e-mail password change requests" received a 28% average click rate [39].

Similarly, other areas of online information can be easily manipulated or can be misleading. For example, the reputation systems of online marketplaces, built by using customer feedback, are most often manipulated by entering the system with multiple names (aliases) [40]. Terrorist groups and criminal gangs use the Internet as a means for planning, communicating, and committing crimes. These crimes can be armed robbery, drug trafficking, and acts of terror [41, 42], with such groups using online messaging systems as safe channels for their covert operations. The digital revolution has greatly simplified the ability to copy and distribute creative works, which has led to increased copyright violations worldwide [43]. As a result, such crimes are challenging to investigate as occurrences are so widespread that it is close to impossible to investigate each one of them.

2.1 Sources of Cybercrimes

There are a variety of sources of cybercrimes that should be considered by investigative solutions and efforts. Sources for cybercrimes can be anything from independent threat actors, hacktivists, to state-sponsored crime. Individual threat actors can include various types of personalities including hackers, script kiddies, e.g., the Mirai-based botnet incident in 2016 by some amateur hackers [44], child pornographers, etc. Hacktivists usually attack to promote some sort of political agenda [2], while individual actors generally commit cybercrime for sport or for financial gain. With state-sponsored cybercrime sources, sophisticated cybercriminals usually supported by governments target other nations to affect the core commercial interests of another country. Attacks originating from nation state are highly targeted and, in most cases, aim to steal intellectual property, get access to business intelligence or military intelligence, and or gain an advantage over a rival nation's technological infrastructure.

The influx of smart connected devices and their potential uses has increased exponentially. Internet users are spending increasing amounts of time online and undertaking a greater range of online and social networking activities, increasing their risk of falling victim to online crime. In most Internet-mediated crimes, the victimization tactics used can be anything including identity theft, masquerading as the victim, harassment, blackmail, bullying, fraud, etc., and the method of communication with the victim can be synchronous or asynchronous. During a synchronize dialogue multiple people are engaged, usually with an expectation of instant feedback amongst the communicants. In this arena, multiple crimes can be committed including harassment, solicitation for prostitution, defraud victims, inciting of violence (terrorism planning), etc. With asynchronous communication, there is an inter delay in the communication medium directed towards an assigned e-mail address. This allows criminals to send malicious communication such as spam for the purpose of opening avenues to exploit a victim or system.

One example of asynchronous communication used for victimization is through the propagation of spam e-mail; distributing unsolicited mail to a large group of people. Spamming can be for nefarious reasons or as a way for a legitimate business to reach a mass audience. An example of spam is an e-mail about unexpected money where a scammer invents convincing and seemingly legitimate reasons to give false hope about offers of money.

The impact of cybercrime on enterprise and society has been discussed in Chap. 1. Notwithstanding the discussion in Chap. 1, cybercrime can change a person's life and cause him/her to fall victim to one of these acts of terror. In instances where spamming is used for nefarious reasons, the effect on the victim can be catastrophic. The evidence of this can be found in the global documentation of several Nigerian 419 spam scams. For example, Janella Spears, an Oregon woman, lost $400,000 to a Nigerian scammer through e-mail spam [45]. As per Janella's story, the scam started with an e-mail which requested a small amount of money with the promise of a return of over 20 million dollars of funds supposedly left behind by her grandfather (J.B. Spears), with whom the family had lost contact over the years. By the time this ordeal was over, she had mortgaged her house, took a lien out on the family car, and ran through her husband's retirement account [45]. Other human faces to this victimization include: in 2007, John Rempel of Leamington, Ontario Canada, lost $150,000 to a Nigeria 419 advance fee scam [46]. These spam scams have not only been limited to monetary losses but also divesting human losses. In November 2003, Leslie Fountain, a senior technician at Anglia Polytechnic University in England, set himself on fire and died later of his injuries after falling victim to an online scam. In 2006, an American living in South Africa hanged himself in Togo after being defrauded by a Ghanaian 419 con man [46]. Similarly, in 2007, a Chinese student at the University of Nottingham killed herself after falling for a lottery scam [46].

Another well-known case is that of Jill Pasovský (February 2003), a 72 year old scam victim from the Czech Republic who shot and killed 50-year old Michael Lekara Wayid, a Nigerian embassy official in Prague, after the Nigerian consular general explained that he could not return $600,000 that Pasovský had lost to a Nigerian scammer [46]. In most of the spam cases resulting in the situations described above, a perpetrator attempts to hide his/her identity while impersonating a higher-authority figure through the phishing e-mail s/he sent to the victim. In web spoofing [47], a potential victim is tricked into entering personal information onto a deceptive website. Likewise, in escrow fraud websites [48], a fake seller creates a dummy online escrow service and then disappears after collecting money from the buyers. There are also potentially more dangerous situations such as predatory and bullying chat conversations.

2.2 Sample Analysis Tools and Techniques in Literature

Installing antivirus, firewalls, network web filters, and intrusion detection systems are not sufficient alone to secure online communication [49]. Moreover, to identify the source of a malicious message, an investigator usually needs to backtrack the Internet Protocol (IP) addresses based on the information in the header of the anonymous message. However, solely tracking the IP address is insufficient to identify the suspect (e.g., the author of an anonymous message) if there are multiple users on the computer that sent out the message, or if the message is sent from a proxy server. In cases of hacked e-mail accounts and compromised computers, the metadata contained in the header cannot be trusted to identify the true originator of the message. Similarly, monitoring chat rooms to detect possible predatory or bullying attacks by entering suspicious chat forums with pseudo-victim Identity (ID) is not a trivial task.

In this context, the forensic analysis of online messages to collect empirical evidence to lawfully prosecute an offender of cybercrime is one way to minimize cybercrimes. The large volume of online messages often contains an enormous amount of forensically relevant information about potential suspects and their illegitimate activities. The existing tools, e.g., Forensic Toolkit (FTK) [50], Encase [51], COPLINK solution suite [52], and Paraben e-mail examiner [53] are some general purpose analysis software and are not designed specifically for analyzing the textual contents of online messages. E-mail Mining Toolkit (EMT) [54], on the other hand, is a free e-mail analysis software that analyzes user behavioral models based on communication patterns. However, the toolkit is limited to analyzing e-mail documents only. The previously mentioned tools do not have the functionality of authorship analysis for resolving authorial disputes.

The challenge is to develop innovative tools and techniques that can be employed to collect forensic evidence by analyzing both the header and the body of an online message—ideally tools and techniques that can work alongside industry standard software. The evidence collected needs to be not only precise but also intuitive, interpretable, and traceable. Header-level information, e.g., IP addresses, hostnames, sender, and recipient addresses contained in an e-mail header, the user ID used in chatting, and the screen names used in web-based communication help reveal information at the user or application level. For instance, the header content extracted from a suspicious e-mail helps reveal who the senders and recipients are and how often they communicate, how many types of communities there are in the dataset, and what are the inter and intra community patterns of communication. There will be more detail on the header level investigation in Chap. 3. The body of a message can be analyzed to collect information about the potential authors and to perceive the underlying semantics of the written text [55].

2.3 Proposed Framework for Cybercrimes Investigation

The proposed framework is designed to automatically perform a multi-stage analysis of suspicious online documents and therefore present the findings with objectivity and intuitiveness. The challenge is to collect evidence that is creditable, intuitive, and can be interpreted by both technical and non-technical professional experts, i.e., law enforcement personnel and jury members. The term online message is used throughout the book to represent the Internet-mediated communication documents, including e-mails, chat logs, blogs, and forum posts. The analysis is applied to the message header as well as the body. A high-level overview of the proposed framework's functionalities is depicted in Fig. 2.2.

Header-content helps reveal some initial clues and preliminary information about the incident that help preparing an initial plan on how to conduct an investigation. For instance, a suspect's e-mail corpus would help revealing e-mail distribution based on sender, recipient, and the time at which a message is sent. Similarly, it would help unfold the social behavior of a suspect within his/her communities and

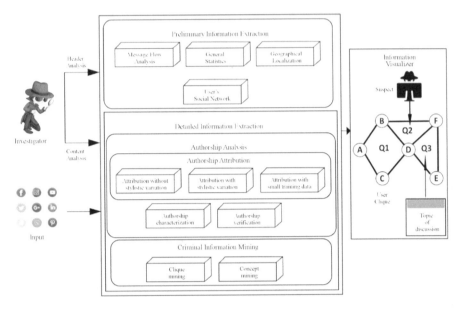

Fig. 2.2 Proposed analysis framework

social groups by applying social networking techniques. The geographic location of a suspect can be achieved by applying geographical localization and map retrieval techniques on the given message collection. Further details on header analysis are discussed in Chap. 3.

The message body is analyzed to collect forensically relevant information about the potential suspects and their activities. The information extracted from the message body is used as internal evidence [56].

Most spam filtering and scanning systems are using topic or content-based classification techniques. Some topic categories with example documents have been used to develop a classification model that is employed for identifying the topic of new messages. Sometimes the task of an investigator is to simply identify the pertinent topics within a large collection of documents without having predefined topics. For this purpose, unsupervised learning technique, called clustering, is applied in this framework.

The presented framework analyzes the message body to (1) collect traces of authorial attributes for addressing the anonymity issue (called authorship analysis) and (2) extract forensically relevant information (called criminal information mining) from the textual content of online messages.

Authorship analysis is the study of linguistic and computational characteristics of the written documents of individuals [57, 58]. Writing styles or specific writing traits extracted from authors' previously written documents can be used to differentiate one person from another [59]. Writing styles are studied mainly in terms of four types of stylometric features: lexical, syntactic, structural, and content-specific. Analytical authorship techniques that have been employed so far include univariate and multivariate statistics [60, 61], machine learning processes such as SVM and decision tree [62, 63], associative classification, as well as frequent-pattern mining [64].

Criminal information mining, on the other hand, is applied to extract knowledge about potential criminal entities and their illegal activities from suspicious online documents. Subsequent chapters will discuss authorship analysis and criminal information mining in detail. Finally, social networking techniques are used to present the extracted information for further investigation.

2.4 Authorship Analysis

When an author writes, s/he uses certain words and combinations of words, whether subconsciously or consciously, creating underlying patterns that characterize that particular author's style of writing. Authorship attribution assumes that each author uses specific words and combinations of words that make his/her writing unique such that features can be extracted from the text that are able to differentiate authors through the use of statistical or machine learning techniques.

Authorship analysis of online documents is more challenging than analyzing traditional documents due to their special characteristics of size and composition [65]. According to [176], "Electronic discourse or online document is neither here nor there, neither pure writing nor pure speech but somewhere in between". Generally, authorship analysis is very useful in circumstances where authorship attribution is uncertain, unknown, or otherwise obfuscated intentionally or unintentionally. Traditionally, authorship analysis has been performed through manual analysis [178]. Unfortunately, part of the challenge in the manual analysis is the increasing use of the electronic medium in communication amongst society. Other challenges with manual analysis are endemic of processing large volumes of electronic or textual data or using traditional stylometric analysis to electronic content.

The traditional literary works such as books and essays are rich sources of learning about the writing style of their authors. Literary works are usually large in size, ranging from few paragraphs to several hundred pages. They are generally well-structured in composition, following definite syntactic and grammatical rules. Most traditional documents are written in a formal way and are intended for a variety of readers. Moreover, the availability of natural language processing tools and techniques make it easy to improve the quality of these documents by removing spelling and idiosyncratic mistakes. The study of stylometric features has long been very successful in resolving ownership disputes over literary and conventional writings

[179]. As a result, it is imperative and justifiable that traditional features used to evaluate authorship for traditional literary works have been operationalized for use with electronic texts. Online documents, on the other hand, are short in size, varying from a few words to a few paragraphs, and often they do not follow definite syntactic and/or grammatical rules; making it hard to learn about the writing habits of their authors from such documents.

Electronic discourses such as e-mail documents have certain properties that help researchers to identify individual writing styles. However, there are multiple roadblocks researchers or investigators must overcome prior to any form of guaranteed automated authorship analysis with an acceptable degree of success. First, electronic data must be contained prior to any investigation or analysis of the data. In cases where an investigator is trying to analyze data from social media data, such as Facebook, data must be collected via an Application Programming Interface (API) or by using an automated crawler. Second, the investigator or researcher must be able to extract necessary stylometric from the data set to begin the analysis required to prove authorship.

Most previous contributions on authorship attribution are applications of text classification techniques [62]. The process starts by identifying a set of a person's writing style features which are relatively common across his/her works. A classifier is trained on the collected writing style features to build a model, which is then used to identify the most plausible author of anonymous documents.

In this book, we study the authorship problem from four main perspectives: (1) Authorship attribution, in which an investigator has a disputed anonymous online message together with some potential suspects. The task of the investigator is to identify the true author of the document in question by analyzing sample documents of potential suspects. (2) Authorship identification with few training samples, in which the available samples of potential suspects are very limited. Further details are presented in Chap. 6. Authorship attribution is studied with and without taking into account stylistic variation of authors in regard to the specific context in which s/he communicates. Authorship attribution is covered in more detail in Chaps. 4 and 5. (3) Authorship characterization. Sometimes an investigator may not have any clue about the suspects and even so, the investigator would like to infer characteristics of the author(s), such as gender, age group, and ethnic group, based on the writing styles in the anonymous messages. See Chap. 7 for more detail on authorship characterization. (4) Authorship verification, confirming whether a given anonymous message is written by a given suspect is the challenge of authorship verification. Some researchers treat verification as a similarity detection problem in which the task is to determine if the two given objects are produced by the same entity, without knowing explicitly about the entity. The challenge is not only to address the authorship problems but to also support the findings with robust forensic evidence. Further detail is provided in Chap. 8.

2.5   Introduction to Criminal Information Mining

In authorship analysis, the task is to extract the content-independent attributes, or stylometric features, e.g., lexical and structural features, from the textual content of documents. In criminal information mining, the task is to analyze the content-specific words of the documents to collect forensically relevant information. The extracted information can be used to answer questions such as; *"What are the pertinent suspicious entities within a document? Are these entities related to each other? What concepts and topics are discussed in the document?"*. Online documents can be analyzed to reveal information about suspicious entities and their malicious activities. Identifying the semantics of the written words by applying contextual analyses and disambiguation techniques will help retrieve malicious documents. Understanding the perceived semantic meaning of suspicious messages is not trivial due to the obfuscation and deception techniques used by perpetrators in their online communication. For instance, the perceived meaning of written words in a malicious discourse is different from their apparent meaning, as the 'street names' used for most illegitimate activities are borrowed from daily conversation, e.g., in e-mails used for drug trafficking the word 'thunder' means heroin and the word 'snow' means cocaine.

There are more than 2300 street terms (used for drugs or drug-related activities) available on the United States Office of National Drug Control Policy.[1] Predictive machine learning methods and natural-language-processing techniques are applied to extract this information. The following section presents state-of-the-art criminal information mining techniques.

2.5.1   Existing Criminal Information Mining Approaches

The textual content of a document can be analyzed to collect forensically relevant information that can be used to answer the following questions: *"Who is the potential author of a text discourse? What are the pertinent suspicious entities mentioned within a document? Are these entities related to each other? What concepts and topics are discussed in the document(s)?"* [55]. Predictive machine learning measures and natural language processing techniques are applied to extract information. Authorship analysis techniques are used to learn about the potential author of an anonymous discourse. Social networking and link analysis techniques [69] are applied to identify a covert association between crime entities. Similarly, topic identification or topic detection is employed to identify the topic or genre of a document [70]. Text summarization methods [71–74] are applied to extract the summary of a potentially large collection of documents. The following section details much of the prevailing scholia on the topic at hand, as well as their proposed methods.

---

[1] http://www.whitehousedrugpolicy.gov

Named Entity Recognition (NER), a branch of natural language processing, is used to identify information associated with an entity such as the name of a person, place, and company; contact information such as phone, e-mail, and URL; or other attributes such as date-of-birth, vehicle number, or assurance number [75]. Chen et al. [41] employed named entity recognition techniques for extracting criminal identities from police narratives and other suspicious online documents. Minkov et al. [76] proposed techniques for extracting a named entity from informal documents (e.g., e-mail messages). While cybercriminals sometimes use identity deception tactics to falsify their true identities, Wang et al. [77] proposed an adaptive detection algorithm for detecting masqueraded criminal identities. Carvalho and Cohen [78] studied techniques for identifying user signatures and the 'reply part' from the e-mail body.

To facilitate the crime investigation process, Chau et al. [69] applied new link analysis techniques to the Tucson police department database to identify a covert association between crime entities. The proposed techniques, including the shortest path algorithm, co-occurrence analysis, and a heuristic approach, were successful in identifying associations and determining their importance. The study [79] applied association rules mining techniques to suspicious websites, called dark web, for identifying online communication between those accused of the 9/11 attacks.

Topic identification, within a corpus of text documents, is the extraction of pertinent content related to a known topic or the topic to be listed [70]. In information retrieval and browsing, topic identification is generally addressed either in a supervised way or in an unsupervised way [80]. In a supervised way, the problem of topic discovery is handled as a text classification or categorization problem [81]. According to this approach, there exist some predefined topic categories with example documents for each category. To infer the topic of an unknown document, a classification model is developed on the given sample documents. Similarly, unsupervised learning or clustering is applied to identify the pertinent groups of objects based on similarity measures.

Pendar [82] applied automatic text categorization techniques on suspicious chat conversation to identify online sexual predators. Each chat session is converted into a vector of attributes using a bag-of-words model. Attributes are the frequencies of word unigrams, bigrams, and trigrams. The words that appear either very rarely or very frequently in the given chat log are deleted. They developed a classification model by applying SVM and k-nearest neighbor (k-NN) classifiers on some previously known predators' chat conversations. The developed model is then employed to identify the predator (or pseudo-predator) communication from a teenager (i.e., a victim) communication. Elnahrawy [83] compared the performance of three classifiers, i.e., Naive Bayes, SVM, and KNN, for automatically monitoring chat conversation following the general text categorization approach. Studies [84–86] focused on the topic identification of chat logs from a list of some predefined topics. Zhang et al. [87] developed text classification techniques for automatic key-phrase extraction in Web documents.

The unsupervised topic identification or topic discovery is achieved by applying content-based clustering. Clustering is used to uncover useful and interesting text patterns in a corpus without any labelled data [76]. Once each document is converted into a term vector, and the pairwise distance between the term vectors is defined, a clustering algorithm is applied to divide the documents into groups. The objective is to group the documents into clusters such that documents in a same cluster are similar and documents in different clusters are dissimilar. Once the documents are clustered, each cluster is labeled with the topic words. The topic words or the cluster label is identified by using different techniques. The simplest way is to identify the words that are found frequently in a cluster, which are divided into two main categories of clustering algorithms: partitioned or hierarchical. In hierarchical clustering, documents are diagramed into a tree-like structure called a dendrogram. Topics at the top level are more general, becoming more specific while descending toward the terminal nodes. The documents associated with each topic is linked to that node.

In [88], the specific attributes of chat users and the relation between users within a chat room are visually displayed. The authors used metaphors for creating visual data portraits of the attributes extracted from chat content and the conversation patterns of users. Examples of attributes include time since initial posting, participation frequency of a user in a chat room or in a topic, and responses to a posting. Bingham et al. [89] developed a chat analysis tool, called ChatTrack, for summarizing and filtering chat logs. A classifier is trained on a set of predefined concepts or topics with sample documents. The classifier then creates a vector of high-frequency words for each topic category. Next, a conceptual profile is created for a selected chat conversation or chat user by training a classifier on the selected chat sessions. The trained classifier is used to create a vector of selected words. Finally, using the cosine similarity measure [90], the similarity between the profile vector and the predefined concept vectors is calculated. There are more than 1565 predefined concepts' hierarchies and their sample documents.

A criminal information-mining framework proposed in [91], was designed by integrating machine learning and data mining techniques such as link analysis, association rule mining, and social network analysis. The developed framework is believed to have the capability of identifying different kinds of crimes. The major focus of this framework is to collect network-level information (i.e., web addresses). The framework can analyze only structured documents such as police narratives. Xiang et al. [92] focused on visualizing crime data for facilitating the work of an investigator. To automatically analyze large archives of online documents, an investigator requires an integrated software tool.

The limitations of most existing criminal information mining techniques are as follows:

1. Forensic tools and techniques, e.g., COPLINK solutions suite, are used to collect network-level information, e.g., URL and hostname, instead of analyzing the textual content of the documents.
2. Most analysis techniques, designed for text classification and clustering, consider only the frequency of words and not their semantics.

3. The proposed approaches focus on structured data, i.e., formal reports, rather than unstructured data such as chat logs and e-mail messages.
4. Most existing forensic tools are either developed for very high-level analysis, e.g., FTK and Encase, or are limited in the application scope. For instance, the E-mail Mining Toolkit and Paraben e-mail examiner do not address the issue of anonymity.

2.5.2 WordNet-Based Criminal Information Mining

This chapter presents a brief overview of WordNet-based criminal information mining approach (full description is presented in Chap. 10). The proposed approach is based on novel text mining and semantic analysis techniques. These techniques are employed to help investigators perform a multi-stage analysis of electronic discourse, including chat logs. The approach takes suspicious chat logs as input, extracts named entities, divides them into different groups, and then retrieves the chat logs of each group for further processing. Keywords and summaries from each chat collection are extracted, which are then processed to extract key concepts representing the topic of the chat log in question. The extracted data involving suspicious groups and their relationships are visualized in a more intuitive fashion. The techniques employed to accomplish the above-mentioned tasks are discussed below.

The widely used Stanford Named Entity Recognizer, called CRFClassifier, has been employed to extract the named entities. The tool is tested on popular corpora such as MUC-6, MUC-7 (Dataset available for download at https://catalog.ldc.upenn.edu/LDC2003T13), and ACE. To identify the relationships between the entities for determining cliques, frequent-pattern mining techniques are applied. Next, two criteria are used to extract the keywords: first, if the word matches with the street term(s) listed in the domain-specific cybercrime taxonomy; second, if the frequency of the word is above the user-defined threshold. The sentences in which one or more keywords appear constitute the summary. The extracted keywords are converted into concepts and the concepts are converted into key concepts and topics by using WordNet. WordNet is a lexical database, described later in this book. The selection of WordNet for the purpose of concept mining and topic mining is based on: (1) Words that are organized into hierarchies of concepts, called synset (synonyms sets); (2) The hierarchies that are based on both similarity and relatedness; (3) Hyponymy, which means that the WordNet synsets are structured in such a way that abstract concepts (called hypernyms) are given at a higher level, while more specific concepts (called hyponyms) are given at a lower level; and (4) a computer-readable database of commonly used words. Complete details of these concepts are presented in Chap. 10.

2.6   WEKA

Waikato Environment for Knowledge Analysis (WEKA) is a collection of state-of-the-art machine learning algorithms and data processing tools used for solving machine learning problems. WEKA [93] has been developed at the University of Waikato in New Zealand. It is written in Java and distributed under the terms of a public license. Most of the WEKA functionality can be used both from within the WEKA toolkit and outside the toolkit, i.e., they can be called from a Java program. WEKA provides extensive support for the whole process of machine learning including preparing data, constructing and evaluating learning algorithms, and visualizing the input data and results of the learning process. WEKA includes methods for most standard machine learning and data mining problems: regression, classification, clustering, association rule mining, and attribute selection.

Classification methods implemented in WEKA [94], namely Ensemble of Nested Dichotomies (END) [95], J48 [96], Radial Basis Function Network (RBFNetwork) [97], naïve Bayes [98], and BayesNet [99] are commonly used for authorship analysis. The decision tree classifier C4.5 implemented in WEKA is denoted as J48. The three widely used clustering algorithms, Expectation-Maximization (EM), k-means, and bisecting k-means, are also implemented in WEKA. The WEKA native data file is the Attribute-Relation File Format (ARFF).

It is an ASCII text file that describes a list of instances sharing a set of attributes. A sample ARFF consists of two sections: the header and the data, as shown in Fig. 2.3.

Fig. 2.3 A sample ARFF file [180]

```
@relation weather

@attribute outlook {sunny, overcast, rainy}
@attribute temperature real
@attribute humidity real
@attribute windy {TRUE, FALSE}
@attribute play {yes, no}

@data
sunny,85,85,FALSE,no
sunny,80,90,TRUE,no
overcast,83,86,FALSE,yes
rainy,70,96,FALSE,yes
rainy,68,80,FALSE,yes
rainy,65,70,TRUE,no
overcast,64,65,TRUE,yes
sunny,72,95,FALSE,no
sunny,69,70,FALSE,yes
rainy,75,80,FALSE,yes
sunny,75,70,TRUE,yes
overcast,72,90,TRUE,yes
overcast,81,75,FALSE,yes
rainy,71,91,TRUE,no
```

The header, called the data declaration section, contains names of attributes followed by their type. The type of an attribute can be numeric (integer or real), nominal, string, or date, as depicted in Fig. 2.3. The data section starts with the reserved word data, preceded by the symbol '@', and is followed by rows of attribute values. The attributes are ordered having a one-to-one association with the attributes defined in the declaration section. Each row represents one instance of the declared attributes. The missing values are denoted by a question mark within the respective position in the row. Values of string and nominal attributes are case sensitive.

Chapter 3
Analyzing Network Level Information

This chapter provides a brief description of the methods employed for collecting initial information about a given suspicious online communication message, including header and network information; and how to forensically analyze the dataset to attain the information that would be necessary to trace back to the source of the crime. The header content and network information are usually the immediate sources for collecting preliminary information about a given collection of suspicious online messages. The header analysis of an e-mail corpus identifying all the senders, the recipients associated with each sender, and the frequency of messages exchanged between users helps an investigator to understand the overall nature of e-mail communication. Electronic messages like e-mails or virtual network data present a potential dataset or a source of evidence containing personal communications, critical business communications, or agreements. When a crime is committed, it is always possible for the perpetrator to manipulate e-mails or any electronic evidence, forging the details to remove relevant evidence or tampering the data to mislead the investigator. Possible manipulation of such evidence may include back-dating, executing time-stamp changes, altering the message sender, recipient, or message content, etc. However, such attempts of manipulation and misleading can be detected by examining the message header. By examining e-mail header and analyzing network information through forensic analysis, investigators can gain valuable insight into the source of a message that is otherwise not traceable through the message body. Investigators can utilize a range of existing algorithms and models and build on leveraging typical forensic planning. Such models focus on what type of information should be collected, ensuring the forensically sound collection and preservation of identified Electronically Stored Information (ESI). By applying these models, it is possible to achieve a full analysis and collect all the relevant information pertaining to the crime. The collected finding is then compiled to reconstruct the whole crime scene, deduct more accurate and logical conclusions [1].

© The Editor(s) (if applicable) and The Author(s), under exclusive license to
Springer Nature Switzerland AG 2020
F. Iqbal et al., *Machine Learning for Authorship Attribution and Cyber Forensics*, International Series on Computer Entertainment and Media Technology, https://doi.org/10.1007/978-3-030-61675-5_3

When conducting a header investigation or analyzing network information to determine a factual electronic footprint about a crime committed, it is important to follow a forensically sound and approved procedure to guarantee the authenticity and validity of evidence in legal proceedings. Lee et al. proposed a model for such an investigation: the Scientific Crime Scene Investigation Model [101]. This model outlines distinct phases: investigation, recognition, identification, individualization, and reconstruction [102]. Another example of such a framework is that of Casey [103]. Casey's framework puts a significant emphasis on processing and analyzing digital evidence using preservation and classification. The first and last stages are the same in both models, and they both explicitly focus on the investigatory processes which are necessary for investigating header information or analyzing networking information. Any investigative framework on header analysis and the analysis of network information should address issues of acquisition, imaging, data protection, extraction, analysis, and reporting or presentation. One particularly curious aspect of digital crime is that the device that is central to the crime may be the tool used to commit the crime or the target of the crime itself—it could even be both. Because of the perverse role that the electronic device plays, any investigation team should ensure that they apply recommended industry standards to accurately collect and preserve digital evidence [104]. A good methodology to adopt in order to ensure a high level of accuracy and preservation would be Ciardhuáin's [105] proposal, which outlines 13 activities that can be used to achieve this objective. Four of these activities are generally accepted and used by other frameworks or models including the Digital Forensics Research Workshop (DFRWS) and Casey's model [105]. The four most generally accepted activities include: searching for and identifying evidence; evidence collection; evidence examination; and the presentation of the hypothesis. The additional activities include awareness of crime scene evidence; prior authorization; planning of investigation direction and tone; notification; transportation; storage; hypothesis; proof/defense, and dissemination of findings. Although the proposed framework is coherent, it is important not to apply any framework without having a broad understanding of forensic analysis. It is essential for analysts to adopt standard operating procedures, apply guidelines, and use a range of analysis techniques in order to complete a successful cybercrime investigation. Figure 3.1 shows several features of e-mail header and body investigation framework. In the remainder of this chapter, the following analyses will be discussed: Classification, Clustering, Temporal analysis, Geographical localization, and other pertinent methods of analysis.

3.1   Statistical Evaluation

A statistical evaluation of mailbox provides information about the flow of e-mails between users including the total number of users (i.e., senders/recipients), the distribution of e-mails per sender, per recipient, per sender-domain, and per recipient-domain, mailing frequency during different parts of the day and the user's average

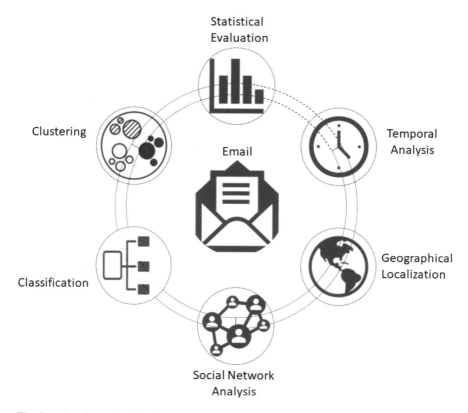

Fig. 3.1 E-mail investigation features

response duration. Statistical evaluation of user's mailbox helps to understand the normal behavior of users and sudden changes in normal behavior reveal the possibility of any suspicious activity. Calculating the average size of a message and its attachment (if there is one) and identifying the format of the message attachment helps create a user's profile. The user profile is used in anomaly detection systems for identifying the suspicious/unusual behavior of users. Figure 3.2 shows a statistical analysis of e-mails. The chart on the top left is showing user e-mails per day for 2 weeks. Investigators can select any specific day from the chart and see e-mail statistics for a particular day. The other charts get updated based on day selection by an investigator.

3.2 Temporal Analysis

In a temporal model, the user network is augmented with time information about e-mails, plotted to show the temporal characteristics of message flow. From this plot, it is easy to identify causality effects between e-mails, for instance, the scenario

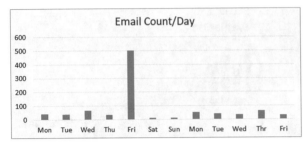

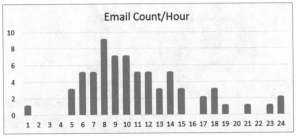

Fig. 3.2 Statistical analysis of e-mail

in which an e-mail is received by a user, who in turn sends another e-mail later. If, for example, both e-mails are classified to the same topic category, e.g., drugs, then by following the chain of the e-mails one can identify the potential collaborators.

3.3 Geographical Localization

The structure of a person's social network, extracted from a dataset, manifests information about his/her behavior with other people, including friends, colleagues, family members, and potential suspicious collaborators. In most investigations, it is important to identify the physical location of the users using e-mail header content. This can be achieved by applying geographical localization and map retrieval techniques to the e-mail addresses. An e-mail header contains the necessary information required to trace its origin. It holds the data footprint or the metadata of each server that the e-mail traveled through, which usually traces back to city/town the e-mail originated from. E-mail header analysis extracts valid evidence from various parts of the header [106]. Generally, investigators begin the e-mail investigation by reviewing header information as it is an important source of evidence for potential e-mail examination. There are several fields in the e-mail header, which can be used to analyze the behavior of the server. These fields can include X-Apparently-To, Delivery To, Return-Path, Received-SPF, Message-ID, MIME-Version, Content-type, X-Mailer, X-Originating-IP & Received, DKIM-Signature. It also contains date and time the message was sent and the filenames of any attachments.

Table 3.1 High-level definition of e-mail header fields [107]

X-Apparently-To-	This value will reveal the recipient's e-mail address. This can be the validation field for checking the e-mail service provider. Generally, this field is referred to as "BCC, CC, or To" and is not restricted to "To"
Delivery To:	This shows the address of the auto-mailer
Return-Path:	This field is used for e-mails that bounce back in case the message cannot be delivered
Received-SPF:	The Sender Policy Framework (SPF) header field shows if the mail server is valid for the sender's e-mail address
Message ID:	This is a globally used unique identifier for each e-mail, which is often comprised of a date/timestamp along with the originating sender domain
MIME-Version:	Multipurpose Internet Mail Extensions (MIME) is an Internet Standard which extends the format of the message
Content-type:	This shows the type of content or format used for the message such as Text or HTML
X-Mailer:	It displays the e-mail client which sent the message
X-Originating-IP:	Identifies the originating IP address of the message, which can help to identify the potential sender
DKIM-Signature:	The Domain Keys Identified Mail (DKIM) signature field contains the digital signature of an e-mail cryptographically signed by the sender's domain to validate the sender's identity

Among all the fields available, message ID plays an important role [107]. While it can often be difficult to determine whether a message is genuine, there can sometimes be useful trails in headers when tactics are less advanced. Table 3.1 defines some common e-mail header fields.

To understand the geographical scope of a cybercrime investigation, it is important to localize the source and destination of the given suspicious messages. This information will help an investigator in collecting additional clues about the potential suspects and their physical locations. This capability can be used to localize information related to potential suspects, e-mail servers, and e-mail flow.

The geographical coordinates of the e-mail server can be identified by employing geographical localization techniques.[1] In situations where localization fails, the server is mapped to a default geographic location in the Atlantic Ocean with the coordinates latitude = 0 and longitude = 0. Once the physical location of each e-mail account is identified, the next step is to display them on the global map.

The proposed framework traces IP address and other details from e-mail header, finds the city level details of the sender, and displays the originating geographic location of the e-mail.

---

[1] http://www.geobytes.com/

3.4 Social Network Analysis

The social network for an e-mail dataset can be depicted as a graph. The structure
of a user's social network, extracted from his/her e-mails manifests a great deal of
information about his/her behavior within his/her community of friends, colleagues,
and family members. This information can be used to answer the following questions [106], (1) How often does the person maintain a relationship with a group of
people, and for how long? (2) Do these people have regular interactions and can
these interactions be distinguished based on roles such as work, friendship, and
family? (3) What type of information a group of people exchanging? For instance,
the analysis of the types of interactions within a criminal network can be used to
discover interesting information about potential suspects and periods during which
their suspicious activities take place. In order to ensure a thorough investigation, it
is important to be able to visualize social networks. Social networks are labeled with
some simple statistics which give information about the flow of messages. In the
graph, called the user model, the nodes denote e-mail users and the edges denote
e-mail traffic. Statistical information computed on a social network is rendered
graphically using features of nodes and links: size, shape, color, thickness, etc. For
instance, the thickness of the links between the nodes denotes the frequency of the
messages sent and the arrow denotes the direction of message flow from sender to
recipient. Nodes associated with users can be replaced with their photos to provide
a more intuitive and elegant representation.

3.5 Classification

Classification or supervised learning is used to identify the class label of an unknown
document based on some previously defined classes. In addition to header-content
analysis, traditional text categorization techniques have also been used [61] for message classification. In general, text classification starts by preprocessing, followed
by classifier training and testing. The validated model is then employed to identify
the class label of the unknown document. The class label is the topic name from a
list of predefined topic categories.

   Classification techniques have been very successful in resolving authorial disputes
over poetic and historic collections. In cybercrime investigation, classification techniques are used for authorship analysis of anonymous messages for identifying perpetrators involved in illegitimate activities. Preprocessing is an essential step in most
text mining processes. Preprocessing starts with message extraction followed by
cleaning, tokenization, stemming, and stop word removal. The available data is often
noisy, containing unwanted and irrelevant information. Consequently, data need to be
converted into a format that can be used by the machine learning process in question.
After extracting the body of an online message (e.g., an e-mail or a chat session), any
text that is written in a language other than English or French (in some cases) is

discarded. E-mail attachments, chains of replies, and (in some cases) HTML tags are also deleted from the e-mails. Java tokenizer API is used to convert each message μ into a set of tokens or words. The different forms of the same word appearing in a message are converted into the root word by applying stemming algorithms, e.g., Porter stemmer [108, 109]. For instance, the words write, wrote, written, and writing is converted into the word 'write'. The list of tokens is then scanned for all-purpose stop words, containing function words (e.g., 'is', 'my', 'yours', and 'below'), short words (e.g., words containing 1–3 characters), punctuation, special characters, and space characters. These words usually do not contribute to the subject matter of a message and are deleted. The actual number of function words varies; [63] lists 150 while [110] mentions 303 function words. The content-specific terms are used for feature extraction. A feature is usually the relative weight calculated for each term. It can be simply the frequency of a term t_j within a message μ_i denoted by $tf_{(i, j)}$; or it can be computed by employing certain functions such as $tf - idf$, described in [111], and is given as:

$$\left(tf - idf\right)_{(i,j)} = tf_{(i,j)} * idf_{(i,j)}$$

where $tf - idf_{(i, j)}$ is the weight of a term t_j within a message μ_i, $tf_{(i, j)}$ is the frequency of a term t_j within message μ_i, $idf_{(i, j)} = \log(N/df_i)$. The inverse document frequency, N is the total number of messages, and df_i is the number of messages where the term t_i appears.

For example, if a message contains 100 words and the word *drug* appears 3 times. The term frequency i.e., tf for the word *drug* is $(3/100) = 0.03$.

Now, if there are one million messages and the word *drug* appear in one hundred of these. Then, the inverse document frequency i.e., idf is calculated as $\log(1,000,000/100) = 4$. So, the $tf - idf$ is the product of these quantities: $0.03 * 4 = 0.12$.

Each message μ is represented as a 'bag of words' using vector space representation [112]. Once all the messages are converted into vectors, normalization is applied to scale down the term frequencies to [0, 1] to avoid overweighing one feature over another. The selected column is scanned for the maximum number and is used to divide all other members of that column. To develop a classification model, the given message collection is divided into two sets: a training set (comprising 2/3 of total messages) and a testing set (comprising 1/3 of total messages). Each message instance of the given sample data carries a class label, representing its topic category.

Common classifiers include decision tree [96], neural networks [111], and Support Vector Machines (SVM) [113] (please see Chap. 3 for details of common classifiers). The validated model is then employed for the classification of a message for which the topic category or author is not known. Usually, the larger the training set, the greater the accuracy of the model. For this purpose, WEKA, a data mining software toolkit was used [94]. The feature vectors are converted into WEKA compatible format: the Attribute-Relation File Format (ARFF). Sometimes, an investigator is asked to analyze a given collection of anonymous documents without any prior knowledge. After preprocessing, the training data is passed to the author classifier and type classifier. Classification is used to identify the type of e-mails and the author of anonymous e-mails.

3.6   Clustering

Clustering is an unsupervised learning process used to retrieve hidden patterns and structures [114] from a dataset without having any previous knowledge. Unlike classification, where an unknown object is assigned to one of the predefined classes, clustering is applied to identify the pertinent groups of objects based on some similarity measure. Clustering is employed for information retrieval [115] and authorship similarity detection [116]. Clustering can be applied to textual content as well as stylometric features.

To initiate the process of investigation the investigator identifies the major topics contained in the given documents. Traditional content-based clustering can be used to first divide the messages into pertinent groups, and then tag each cluster with the most frequent words. In this framework, three clustering algorithms are used: Expectation Maximization (EM), k-means, and bisecting k-means. Once the clusters are obtained, each cluster is tagged with the high-frequency words found in the respective cluster. The clusters can be used for document retrieval by matching the given keywords with the cluster labels. The matched clusters are retrieved in the order of relevance to the search criterion (query content).

Chapter 4
Authorship Analysis Approaches

This chapter presents an overview of authorship analysis from multiple standpoints. It includes historical perspective, description of stylometric features, and authorship analysis techniques and their limitations.

4.1 Historical Perspective

The origin of authorship attribution or identification dates back to the eighteenth century when English logician Augustus de Morgan suggested that authorship might be settled by determining whether one text contained significantly longer words than another. Authorship attribution or identification ascertains the probability of a written piece having been produced for evaluation by a particular author by examining previously written work by the same author to determine or attribute the submitted piece to that author. Generally, this is applied to an anonymous document to determine the likelihood of a specific author having written the document by looking for the presence of data that may be similar to those in previous documents identified as having been written by this author. The problem can be looked at either by using a statistical hypothesis test or from a classification perspective. If we look at this from a classification perspective, then we are looking at the ability to identify feature sets that remain relatively constant across a large number of written documents belonging to the same author. Once a feature set has been selected, a piece of written work can then be represented by an n-dimensional vector, where n is the total number of features selected. With a set of categorized vectors, different analytical techniques can be applied to determine the category of a new vector created based on a new piece of writing. Hence, the features set, and the analytical techniques may significantly affect the performance of authorship identification.

F. Iqbal et al., *Machine Learning for Authorship Attribution and Cyber Forensics*, International Series on Computer Entertainment and Media Technology, https://doi.org/10.1007/978-3-030-61675-5_4

4.2   Online Anonymity and Authorship Analysis

The growing trend and integration of the Internet of Things and related applications continue to provide multiple avenues for electronic information dissemination across various media. As well as its ease of use and efficiency, one of the common attributes of online communication is the ability of the individual to create a totally different persona. As society drives towards exclusively using digital communication media, there is an opportunity for people with malicious intent to exploit online communication for criminal or unethical activities. Such activities may result in the distribution of unsolicited or inappropriate content e.g., spam e-mails, harassment, offensive content, or even illegal materials such as distributing child pornography materials or terrorist recruitment materials. With online communication, there is a risk of the infringement of content under copyright. Furthermore, internet-enabled devices can also be used as a means of communication for groups intending to cause harm or disturbance. A certain level of anonymity is created as a result of people not being forced, in many cases, to provide real and verifiable identity information such as their name, age, gender, and address. This allows the creator or sender of malicious electronic communication to forge a new persona which makes the identification of the sender more challenging during a criminal investigation. There is an increasing need for efficiently automated methodologies to easily identify and trace authors of malicious content.

Authorship classification methodologies help to identify the culprits of crimes including the misuse of online communication tools for sending blackmail, harassment, or spam e-mail. Many authors of such materials use all available means in an attempt to hide their true identities, misleading investigators, and attempting to avoid detection. A criminal can hide his or her true identity by impersonating another sender, using encryption technologies, or routing electronic communication through an anonymous server using pseudonyms while distributing criminal content through different anonymous channels. Most authorship classification approaches leverage machine learning methods, information retrieval, and natural language processing in order to achieve authorship identification. An effective classification model can ascertain whether a given author, for whom we have an existing corpus of writing samples, is the same author as the one who wrote a document that was used as a piece of evidence. This is useful during an investigation to determine the proprietor of a crime.

4.3   Stylometric Features

Several features are critical in order to extract a unique writing style from a number of online messages, such as lexical features, content-free features, syntactic features, structural features, and content-specific features.

As the nature of most crimes and the tools used to commit crimes have changed, traditional tools and techniques may no longer be applicable in prosecuting cyber-criminals in a court of law. The statistical study of stylometric features, called sty-lometry, shows that individuals can be identified by their relatively consistent writing styles. The writing style of an individual is defined in terms of word usage, selection of special characters, the composition of sentences and paragraphs, and organization of sentences into paragraphs and paragraphs into documents. Rudman has identified more than 1000 stylometric features in his study [117]. But no feature set exists that is sufficiently optimized and equally applicable to all people and in all domains. Previous authorship studies [57, 58, 65, 116] contain lexical, syntactic, structural, and content-specific features. Other features studied in authorship litera-ture include idiosyncrasies [110], n-grams (e.g., bigrams and trigrams), and fre-quency of part-of-speech tags [118]. A brief description and the relative discriminating capability of the main feature types are given below:

- *Lexical features* are used to learn about an individual's preferred use of isolated characters and words. These include the frequency of individual letters of the alphabet (26 letters of English), the total number of uppercase letters, capital let-ters used at the beginning of sentences, the average number of characters per word, and the average number of characters per sentence. The use of such fea-tures indicates an individual's preference for certain special characters or sym-bols or the preferred choice of selecting certain units. For instance, some people prefer to use the '$' symbol instead of the word 'dollar', '%' for 'percent', and '#' instead of writing the word 'number.' Word-based lexical features, including word length distribution, words per sentence, and vocabulary richness, were very effective in earlier authorship studies [119–121]. Recent studies on e-mail authorship analysis [63, 65] indicate that word-based stylometry such as vocabu-lary richness is not very effective for two reasons. First, e-mail messages and online documents are very short compared to literary and poetry works. Second, word-oriented features are mostly context-dependent and can be consciously controlled by people.
- *Syntactic features* include content-independent all-purpose function words, e.g., 'though', 'where', and 'your'; punctuation, e.g., '!' and ':'; and part-of-speech tags. Mosteller and Wallace [122] were the first to show the effectiveness of the function words in addressing the issue of Federalist Papers [120]. Burrows [58] used 30–50 typical function words for authorship attribution. Subsequent studies [57] have validated the discriminating power of punctuation and function words. Zheng et al. [63] used more than 150 function words. Stamatatos et al. [123] used frequencies of part-of-speech tags, passive account, and nominalization count for authorship analysis and document genre identification.
- *Structural features* are helpful to learn how an individual organizes the layout and structure of his/her documents. For instance, how are sentences organized within paragraphs, and paragraphs within documents? Structural features were first suggested by de Vel et al. [65, 124] for e-mail authorship attribution. In addi-tion to the general structural features, they used features specific to e-mails such

as the presence/absence of greetings and farewell remarks and their position within the e-mail body. Moreover, within e-mails, some people use the first/last name as a signature while others prefer to include their job title and mailing address as well. Malicious e-mails often contain no signatures and, in some cases, may contain fake signatures.

- *Content-specific features* are used to characterize certain activities, discussion forums, or interest groups by a few keywords or terms. For instance, people involved in cybercrimes (spamming, phishing, and intellectual property theft) commonly use "street words" such as 'sexy', 'snow', 'download', 'click here', and 'safe', etc. Usually, term taxonomy built for one domain is not applicable in another domain and can even vary from person to person in the same domain. Zheng et al. [63] used around 11 keywords (such as 'sexy', 'for sale', and 'obo') from the cybercrime taxonomy in authorship analysis experimentations. A more comprehensive list of stylistic features including idiosyncratic features was used in [110].

- *Idiosyncratic features* include common spelling mistakes, e.g., transcribing 'f' instead of 'ph' (as in the word phishing) and grammatical mistakes, e.g., writing sentences with the incorrect form of verbs. The list of such characteristics varies from person to person and is difficult to control. Gamon [125] achieved high accuracy by combining certain features including part-of-speech trigrams, function word frequencies, and features derived from semantic graphs.

4.4   Authorship Analysis Methods

Authorship attribution of online messages has become a forthcoming research area. It overlaps with many research areas such as statistical analysis, machine learning, information retrieval, and natural language processing. The authorship attribution problem started as the most basic problem of the author identification of texts with the unknown author, but the authorship attribution problem has been integrated and extended into several fields e.g., forensic analysis, electronic commerce, etc. This extended version of the author attribution problem, where the search space is very large and there is usually no clue about the potential author, has been defined as a needle-in-a-haystack problem.

Authorship analysis techniques are generally employed to detect the most likely author of a sample of writing, amongst a lineup of potential culprits during a cyber forensic investigation, in an attempt to prosecute or exonerate an accused person. It is apparent that with the increase of well-defined cyber laws and the increasing growth of cyber-crimes such as harassment, fraud, cyberbullying, etc., it is imperative for investigators to ascribe to authorship classification techniques which will allow grouping, analysis, and which will use of the best classification methodology in the identification of a criminal from a suspect population.

Authorship is studied using various computational and analysis methods. These methods including statistical and cluster analysis, machine learning, frequent pattern mining, and associative classification, are discussed in detail in this book. The classifiers, commonly used in these studies, fall into three main categories: (1) probabilistic classifiers, e.g., Bayesian classifiers [95] and its variants; (2) decision tree [93], e.g., C4.5 and J48; and (3) support vector machine [107] and its variants, e.g., Ensemble SVM.

4.4.1 Statistical Analysis Methods

The authorship problem is studied using CUSUM and cluster analysis statistical analysis methods. The Cumulative Sum Control Chart, the CUSUM, is a sequential analysis technique used for monitoring small changes in the process means. As its name implies, CUSUM involves the calculation of a cumulative sum. The CUSUM statistics procedure has been used in authorship analysis. The main concept of this process is to create the cumulative sum of the deviations of the measured variable and plot that in a graph to compare the writing samples of different authors with the submitted written sample to be classified. This technique has proven to be successful and it has been used as a forensic tool to help analysts perform authorship analysis. Unfortunately, CUSUM analysis lacks stability when testing multiple topics, making it unreliable. Univariate methods such a CUSUM are known to have a constraint which prevents them from dealing with more than two features simultaneously. The presence of these constraints makes it necessary to seek other ways of dealing with multiple features. By applying multivariate approaches such as principal component analysis (PCA), it is possible to deal with multiple features and the frequency of function words. PCA can merge many measures and is able to project them into a graph. The geographic distance on the graph shows how similar the author styles are.

Cluster analysis is a multivariate method for solving classification problems. It sorts objects (people, things, events, etc.) into groups or clusters in such a way that members of the same cluster are more similar to each other than those in other clusters. Mutually supportive results found by a variety of multivariate methods have further validated the usefulness of multivariate methods.

4.4.2 Machine Learning Methods

The advent of powerful computers instigated the extensive use of machine learning techniques in authorship analysis.

(a) **Naïve Bayes classifier:** The Naïve Bayes classification method is based on probability theory. In the literature, the Bayes theorem has played a critical role in probabilistic learning and classification. Given no information about an item, it uses the prior probability of each category to deal with the problem of authorship. The Naïve Bayes (NB) classifier involves the popular identification or attribution technique characterizing the stationary distribution of words or letters. Systematic work carried out with the NB classifier can provide solid evidence, making this approach a solid one that has brought a lot to the field.

(b) **Feed-forward neural network:** An artificial neural network where connections between the nodes do *not* form a directed cycle is called a feed-forward neural network. This is different from traditional networks. The feed-forward neural network was the first and simplest type of artificial neural network devised. In this network, information moves in only one direction, forward, from the input nodes, through the hidden nodes (if any), and to the output nodes. There are no cycles or loops in the network.

(c) **Radial basis function network:** A radial basis function network is an artificial neural network that uses radial basis functions as activation functions. The output of the network is a linear combination of radial basis functions of the inputs and neuron parameters. Radial basis function networks are used for function approximation, time series prediction, and system control. Radial basis function (RBF) networks were applied to investigate the extent of Shakespeare's collaboration with his contemporary, John Fletcher, on various plays.

(d) **Support Vector Machines:** In machine learning, support vector machines (SVMs) are supervised learning models with associated learning algorithms that analyze data and recognize patterns. They are used for classification and regression analysis. The basic SVM takes a set of input data and predicts, for each given input, which of two possible classes forms the output, making it a non-probabilistic binary linear classifier. Diederich et al. introduced SVM to this field and carried out experiments [126] to identify the seven target authors writing from a set of 2652 newspaper articles written by several authors covering three topic areas. In this study, the SVM method achieved 60–80% accuracy in detecting the target author. The performance of authorship identification for SVM is directly affected by parameters such as the number of authors to be identified and the size of training data to train the model. As a popular classification method, SVM was applied by Olivier [62] over a set of structural and stylistic features for e-mail authorship attribution. After performing several experiments, Olivier concluded that the accuracy of classification is directly proportional to the size of the training dataset and average document length whereas it is inversely proportional to the number of authors. This explains the decline in classification accuracy that is seen when processing documents with e-mail-like characteristics. It was further found that the performance of SVM was diminished when the number of function words used increased from 122 to 320, contradicting the tenet that SVM supports high dimensionality and leading to the conclusion that increasing the number of features does not improve accuracy. However, it has been proposed by Iqbal et al. [59] that identifying key combinations of features that are able to differentiate between the writing styles of various suspects and removing useless or noisy features can improve accuracy. Previous authorship attribution techniques also suffered from the challenge of considering too many features, making it difficult to isolate the right feature sets to use for any given e-mail set. Olivier [62] showed that adding meaningless features may decrease the accuracy of classification when a classifier captures these features as noise. Using common or otherwise weak features for classification also damages the justification of evidence for corroborating the finding, thus creating a legal problem from a technical one.

4.4.3 Classification Method Fundamentals

This section outlines existing authorship classification techniques. Most classification approaches can be grouped into either statistical univariate methods or machine learning methods/techniques. Unlike authorship verification, which is studied as a one-class and two-class classification problem, modern authorship identification can be approached as a multi-class classification problem. It can be better understood by reading Stamatato's survey. There is no single standard or predefined set of features that is the best at differentiating the writing style of individual writers, but some studies have identified the most representative common features in terms of correctly classifying anonymous or disputed texts. Punctuation and n-gram features have been shown to be highly representative on their own but combining these features was discovered to be even more effective. The relative preference for using certain words over others along with their associations is another highly representative feature. Vocabulary richness, fluency in the language, and the grammatical and structural preferences of individuals are among the most important writing style manifestations.

Most classification methods require feature selection as a key step in ensuring accuracy. Authorship analysis has been quite successful in resolving authorship identification disputes over various types of writing. However, e-mail authorship attribution poses special challenges due to the fact that the size, vocabulary, and composition of e-mails differ considerably from literary works. Literary documents are usually large in size; they have a definite syntactic and semantic structure. In contrast, e-mails are short and usually do not follow well-defined syntactic or grammar rules. Thus, it is harder to model the writing patterns of the authors of e-mails. Various approaches to authorship classification propose solutions to deal with this challenge.

In the literature, authorship problem has been discussed from the following three perspectives (1) authorship attribution, (2) authorship characterization or authorship profiling, and (3) authorship similarity detection or authorship verification. Details are presented in the subsequent sections.

Later in this book, we discuss a special case of authorship attribution where training samples of candidate suspects are very few and are not sufficient to train a classifier. Yet there is another special case where we focus the occasional variation in the writing style of individuals subject to the contexts in which they write (See Chaps. 5 and 6).

4.5 Authorship Attribution

The goal of authorship attribution or authorship identification in the context of online documents is to identify the true author of a disputed anonymous document. In forensic science, an individual can be uniquely identified by his/her fingerprint. Likewise, in cyber forensics, an investigator seeks to identify specific writing styles, called wordprint or Writeprint, of potential suspects, and then use them to develop a model. The Writeprint of a suspect is extracted from her previously written documents. The model is applied to the disputed document to identify its true author among the suspects. In forensic analysis, the investigator is required to support her findings by convincing arguments in a court of law.

In the literature, authorship identification is considered as a text categorization or text classification problem. The process starts with data cleaning followed by feature extraction and normalization. Each document of a suspect is converted into a feature vector using a vector space model representation [90]; the suspect represents the class label. The extracted features are bifurcated into two groups, training, and testing sets. The training set is used to develop a classification model while the testing set is used to validate the developed model by assuming the class labels are not known. Common classifiers include decision tree [96], neural networks [127], and SVM [111]. If the error approximation is below a certain acceptable threshold, the model is employed.

Generally, a disputed anonymous document is preprocessed and converted into a feature vector in a manner similar to the one adopted for known documents. Using the developed model, the conceivable class label of the unseen document is identified. The class label indicates the author of the document in question. Usually, the larger the training set, the better the accuracy of the model. The accuracy of the model is gauged by employing the popular functions called *precision* and *recall*, described in [122]. The difference between traditional text classification and authorship classification is that in text categorization syntactic features, e.g., punctuation, all-purpose stop words, and spaces, are dropped and the features list includes topic-dependent words, while in authorship problems, topic words or content-dependent words are removed, and the features are calculated in terms of style markers or syntactic features. Similarly, in text categorization problems the class label is the topic title among the predefined document categories, while in authorship attribution the class label is the author of the document.

Most authorship attribution studies differ in terms of the stylometric features used and the type of classifiers employed. For instance, Teng et al. [49] and de Vel [62] applied the SVM classification model over a set of stylistic and structural features for e-mail authorship attribution. de Vel et al. [65] and Corney et al. [124] applied SVM on an e-mail dataset and discovered the usefulness of structural features for e-mail authorship attribution. They have also studied the effects of varying the number of authors and sample size on the attribution accuracy. Zheng et al. [63, 116] and Li et al. [115] use a comprehensive set of lexical, syntactic, and structural features including 10–11 content-specific keywords. They used three classifiers including C4.5, neural networks, and SVM for authorship identification of online documents written in English and Chinese. Van Halteren [128] used a set of linguistic features for the authorship attribution of student's essays. In [129], different classifiers were evaluated for the authorship identification of chat logs. Zhao and Zobel [130] have studied the effectiveness of function words in authorship problems by applying different classifiers. de Vel [65] found that by increasing the number of function words from 122 to 320, the performance of SVM drops, due to the scalability problem of SVM. This result also illustrates that adding more features does not necessarily improve accuracy. In contrast, the focus is to identify the combinations of key features that can differentiate the writing styles of different suspects and filter out the useless features that do not contribute towards authorship identification.

Some research proposals [65, 131] have recognized the contextual and temporal change in the writing style of a person, although most choose to ignore such variations and focus on obtaining the permanent writing traits of an individual. Therefore, they extract stylometric features from the entire sample dataset of a suspect, disregarding the context and the type of recipient of a message. In fact, the writing style of an individual varies from recipient to recipient and evolves with the passage of time and with the context in which a message is written [65]. Style variation is a factor of the commonly used four types of writing style features. For example, the change in the topic of an online message is indicated by the relative composition of words and phrases. Official messages may contain more formal words and phrases that may result in an increased value of vocabulary richness. Similarly, syntactical features, including punctuation, hyphenation, and the distribution of function words, are usually more frequent in an online text written to senior management within a company. Moreover, the ratio of spelling and grammatical mistakes is usually higher in the electronic discourse sent to a friend than to a co-worker. Malicious messages, on the other hand, may contain more fancy and charming words that are appealing and attractive to the target victims. Words like 'congratulations!', 'hurry up', 'free download' and 'obo' are commonly found in spamming messages. The analytical techniques employed over such intermingled writing samples would produce misleading results.

4.6 Authorship Characterization

Authorship characterization [56, 124] is applied to collect sociolinguistic attributes such as gender, age, occupation, and educational level, of the potential author of an anonymous document. In the literature, authorship characterization is addressed as a text classification problem. Generally, a classification model is developed by using the textual documents previously written by the sample population. The developed model is applied to the anonymous document to infer the sociolinguistic characteristics of the potential anonymous author. Corney et al. [124], Koppel et al. [56, 132], and Argamon et al. [55] studied the effects of gender-preferential attributes on authorship analysis. Other profiling studies have discussed the educational level [124], age, language background [56], and so on. To address the same issue in the context of chat datasets, some techniques have been proposed in [129] for predicting the potential author of a chat conversation. The proposed technique is employed to collect sociolinguistic and demographic information such as the gender, age, and occupation of the writer of an anonymous chat segment. Abbasi and Chen [40] applied similarity detection techniques on customer feedback to identify fake entities in the online marketplace. In [56, 124], authorship profiling was applied to collect the demographic and sociolinguistic attributes of the potential author of a disputed document.

Existing characterization studies vary in terms of the type of classifiers used, the dimension of characteristics inferred, and the nature of documents analyzed. For instance, Corney et al. [124] and de Vel et al. [133] used SVM to infer the gender, educational level, and language background of an e-mail dataset. Koppel et al. [56] applied a Bayesian regression function to predict the gender, age, and native language of the perceived author of an anonymous text. Most characterization studies are based on classifiers, which are not suitable for use in forensic analysis due to some limitations. There is a discussion about the limitations of characterization studies earlier in this chapter. A frequent-pattern based Writeprint extraction, representing the unique writing style of an individual, has been proposed. Unlike traditional techniques, frequent-pattern based extraction does not require large training data for producing trustable results. Similarly, it can be applied to most text discourses, although the current study is focused on a blog dataset.

4.7   Authorship Verification

Unlike authorship attribution and authorship characterization, where the problem is clearly defined, there is no consensus on how to precisely define the problem in authorship verification studies. Some studies, e.g., [62, 110] have considered it as a similarity detection problem: to determine whether two given objects are produced by the same entity or not, without knowing the actual entity. Internet-based reputation systems, used in online markets, are manipulated by using multiple aliases of the same individual. Novak et al. [134] proposed a new algorithm to identify when two aliases belong to the same individual while preserving privacy. His technique has been successfully applied to postings of different bulletin boards, achieving more than 90% accuracy. To address the same issue of similarity detection, Abbasi and Chen [40, 110] proposed a novel technique called Writeprints for authorship identification and similarity detection. They used an extended feature list including idiosyncratic features in their experimentations. In similarity detection, they took an anonymous entity, compared it with all other entities, and then calculated a score. If the score is above a certain predefined value, the entity in hand is clustered with the matched entity.

Following the same notion of verification, Halteren [128] proposed a different approach called linguistic profiling. In this study, he proposed some distance and scoring functions for creating profiles for a group of example data. The average feature counts for each author were compared with a general stylistic profile built from the training samples of widely selected authors. The study focused on detecting similarities between student essays for plagiarism and identity theft. The more common notion of authorship verification involves confirming whether or not the suspect is the author of a disputed anonymous text. Some studies address authorship verification as a one-class classification problem (e.g., [130] and [127]) while others (e.g., [56] and [132]) as a two-class text classification problem. For instance, Manevitz et al. [127] investigated the problem as follows: Given a disputed document together with sample documents of the potential suspect, the task is to verify whether or not a given document is written by the suspect in question. Documents written by the sample population are labeled as 'outliers' in their study. A classification model is developed using the stylometric features extracted from the collected documents. The built model is applied to identify the class label of the given anonymous document.

A slightly modified version of the one-class approach called the 'imposter' method is the two-class problem proposed by Koppel et al. [132]. According to this study, the known documents of the potential suspect are labeled 'S' and those of the sample population are labeled as 'imposter'. A classification model is developed using the stylometric features extracted from these documents. The anonymous document is divided into different chunks and each chunk is given to the model to predict its class. The method fails to work if the documents of the 'imposter' and the suspect are closely similar. An alternative approach would be to train one model for S and for not-S and then employ a trained model to determine the degree of dissimilarity between them [56]. In this study, the authors employed the traditional tenfold cross-validation approach. If the validation accuracy is high, it is concluded that S did not write the document in question. Otherwise, the model fails to assign a class label.

A relatively new approach, called 'unmasking' [56], is an extension of the 'imposter' method. In this study, the authors attempted to quantify the dissimilarity between the documents of the suspect and that of the 'imposter.' The experimental results reported indicate that the method is suitable and capable of generating reliable results in situations where the document in question is at least 5000 words long.

4.8 Limitations of Existing Authorship Techniques

Most of the existing authorship analysis techniques are primarily based on some commonly used classifiers. As mentioned, these classifiers can be broadly divided into three main categories: probabilistic [98], decision tree [135, 136], and SVM [137]. Each of these classifiers has its own limitations in terms of classification accuracy, scalability, and interpretability. Probabilistic Naive Bayes classifiers and batch linear classifiers (Rocchio) [138] seem to be the least effective of the learning-based classifiers, while SVM appears to be the most effective in terms of classification accuracy [81].

While building a decision tree a decision node is selected by simply considering the local information of one attribute, it fails to capture the combined effect of several features. In contrast, SVM avoids such a problem by considering all features when a hyperplane is created. However, SVM is like a black-box function that takes some input, i.e., a malicious message, and provides an output, i.e., the author. It fails to provide an intuitive explanation of how it arrives at a certain conclusion. Therefore, SVM may not be the best choice in the context of the forensic investigation, where collecting credible evidence is one of the primary objectives.

Most classifiers require a sufficiently large sample of training data in order to produce an acceptable level of accuracy in terms of classification. The collected training samples from the suspects in criminal investigation cases are not always enough to train a classifier. Therefore, an approach needs to be designed that can work even with a small amount of training data. Most authorship techniques that are successful in resolving authorial disputes of structured documents, e.g., books and formal reports, may not produce reliable results in the context of online messages due to their small size and casual content.

To overcome the limitations of existing authorship techniques, a novel approach has been developed for authorship analysis (Illustrated in Chap. 5). In this method, a unique Writeprint is created for each suspect based on his or her previously written documents. The concept of the Writeprint is based on the idea of the frequent pattern [139], a data mining technique. Frequent-pattern mining has been very successful in finding interesting patterns in large archives of documents analyzed for the identification of customer purchasing habits, cataloging objects in large superstores, intrusion detection systems, and traffic classification.

Chapter 5
Writeprint Mining For Authorship Attribution

This chapter presents a novel approach to frequent-pattern based Writeprint creation, and addresses two authorship problems: authorship attribution in the usual way (disregarding stylistic variation), and authorship attribution by focusing on stylistic variations. Stylistic variation is the occasional change in the writing features of an individual, with respect to the type of recipient and the topic of a message. The authorship methods proposed in this chapter and in the following chapters are applicable to different types of online messages; however, for the purposes of experimentation, an e-mail corpus has been used in this chapter, to demonstrate the efficacy of said methods.

The problem of authorship attribution in the context of online messages can be described as follows: a cyber forensic investigator wants to determine the author of a given malicious e-mail ω and has to prove that the author is likely to be one of the suspects $\{S_1, \cdots, S_n\}$. The problem is identifying the most plausible author from the suspects $\{S_1, \cdots, S_n\}$ and gathering convincing evidence to support such a finding in a court of law. The problem of authorship identification in the context of e-mail forensics is distinct from traditional authorship problems in two ways. Firstly, the number of potential suspects is larger and the (usually confiscated) documents that are available to the investigator (e.g., e-mails) are greater in number. Secondly, by assumption, the true author should certainly be one of the suspects. The problem of authorship analysis becomes more challenging by taking into consideration the occasional variation in the writing style of the same person. The authorship attribution studies used to inform this chapter [63, 65] discuss contextual and temporal variation in people's writing styles, but none of these studies propose methods for capturing stylistic variation. The writing style of a suspect may change either due to a change in the context (or the topic of discussion in an e-mail) or the type of

Some contents in this chapter are developed based on the concepts discussed [64] F. Iqbal, R. Hadjidj, B. C. M. Fung, and M. Debbabi, "A novel approach of mining write-prints for authorship attribution in e-mail forensics," Digit. Investig., vol. 5, pp. S42–S51, 2008.

F. Iqbal et al., *Machine Learning for Authorship Attribution and Cyber Forensics*, International Series on Computer Entertainment and Media Technology, https://doi.org/10.1007/978-3-030-61675-5_5

recipient [65]. Employing analytical techniques over the entire collection of an author's writing samples without considering the issue of stylistic variation (the term coined for the first time in the current study) would produce misleading results [64].

In this chapter, a novel approach of extracting a frequent-pattern based "Writeprint" has been presented to address the attribution problem in the usual way, as depicted in Fig. 5.1. Then, the proposed approach is extended to address the same problem of authorship attribution with a stylistic variation or stylistic inconsistency, as shown in Fig. 5.2. The use of fingerprinting techniques for identifying a potential suspect in a traditional criminal investigation process is not applicable to the digital world.

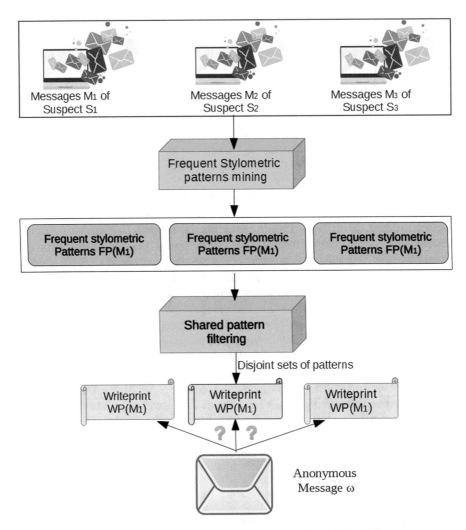

Fig. 5.1 AuthorMiner1: Authorship identification without stylistic variation [64]

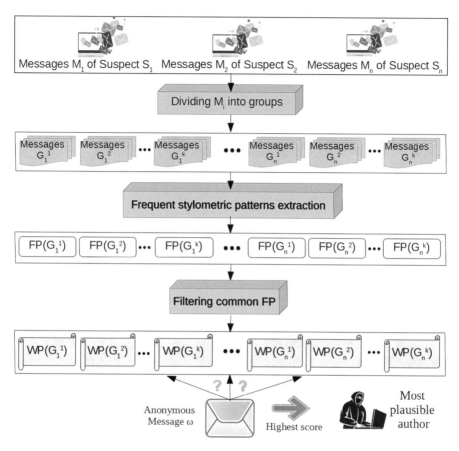

Fig. 5.2 AuthorMiner2: Authorship identification with stylistic variation [140]

However, authorship studies [58, 120] suggest that people usually leave traces of their personality in their written work. Therefore, in cyber forensics, an investigator seeks to identify the "Writeprint" of an individual from his/her e-mail messages and use it for authorship attribution. The key question is: What exactly are the patterns that can represent the Writeprint of an individual?

Our insight is that the Writeprint of an individual is the combination of features that occur frequently in his/her written e-mail messages. The commonly used features are lexical, syntactical, structural, and content-specific attributes. By matching the Writeprint with the malicious e-mail, the true author can be identified. Most importantly, the matched Writeprint should provide credible evidence for supporting the conclusion. The research community [49, 62, 63] has devoted a lot of effort to studying stylistic and structural features individually, but few have studied the combinations of features that form a Writeprint and addressed the issue of evidence gathering.

Figure 5.1 depicts an overview of the proposed method, called AuthorMiner1, for addressing the usual attribution problem. First, the set of frequent patterns are extracted independently from the e-mail messages M_i written by suspect S_i. Though the set of frequent patterns captures the writing style of a suspect S_i, it is inappropriate to use all the frequent patterns to form the Writeprint of a suspect S_i because another suspect, say S_j, may share some common writing patterns with S_i. Therefore, it is crucial to filter out the common frequent patterns and identify the unique patterns that can differentiate the writing style of a suspect from that of others. These unique patterns form the Writeprint of a suspect.

To address the attribution problem with stylistic variation, an extended version of the AuthorMiner1 has been developed, called AuthorMiner2. An overview of AuthorMiner2 is shown in Fig. 5.2 and is outlined in Algorithm 5.2. With AuthorMiner2, the first step is as follows: each message collection M_i of a suspect S_i is divided into different groups $\{G_i^1, \cdots, G_i^k\}$. Next, frequent stylometric patterns $FP(G_i^g)$ from each group G_i^g are extracted.

Following this, the frequent patterns shared between two or more groups across all the suspects are deleted. The remaining frequent stylometric patterns form the sub-Writeprint of each group G_i^g, denoted by $WP(G_i^g)$. In the fourth step, the most plausible author S_a of ω is identified by comparing every extracted Writeprint $WP(G_i^g)$ with ω.

This approach to authorship identification has benefits that are not found in most existing works.

- **Justifiable evidence**: The Writeprint, represented as a set of unique patterns, is extracted from the sample documents of a particular suspect. The proposed method guarantees that the identified patterns are frequent in the documents of one suspect only and not frequent in the documents written by others. It would be difficult for an accused suspect to deny the validity of the findings. The results obtained are traceable, justifiable, and can be presented quantitatively with statistical support. Traditional authorship identification methods, such as SVM and neural networks, do not have the same benefits.
- **Flexible writing styles:** The frequent pattern-mining technique can adopt all four types of commonly used writing style features (described in sections below). The technique is much more flexible than the traditional decision tree, which primarily rely on the nodes at the top of the tree to differentiate the writing styles of all suspects. This flexibility is important for determining the combined effect of different features.
- **Features optimization:** Unlike traditional approaches, where it is hard to determine the contribution of each feature in the authorship attribution process [65], the proposed frequent pattern-mining technique is based on distinctive patterns and the combination of features. The support associated with each pattern in the Writeprint set determines the contribution of each pattern.
- **Capturing inconsistent stylistics:** Analyses show that the writing style of a person is not usually consistent and may change depending on the recipients and the context of the message. The proposed algorithm in this chapter can capture the

sub-stylistic attributes of an individual by employing the idea of frequent patterns. Further experimental results suggest that the identification of sub-Writeprints can improve the accuracy of authorship identification. Most importantly, the sub-Writeprint reveals the fine-grained writing styles of an individual that can be valuable information for an investigator.

5.1 Authorship Attribution Problem

The problem of authorship attribution is divided into two subproblems. The first is the traditional authorship attribution problem in which the occasional change in the writing style of a person is ignored. Meanwhile, investigators try to extract the stylometric features from the entire message collection of a suspect. The second subproblem takes the style inconsistency or stylistic variation of a suspect into consideration and proposes methods for dividing the messages of each suspect into different groups to capture the stylistic variation prior to applying the authorship identification process.

5.1.1 Attribution Without Stylistic Variation

Investigators have access to the training samples of the suspects. In a real-life investigation, sample text messages can be obtained from the suspects' e-mail archives and chat logs on the seized personal computer, or from the e-mail service provider with warrants. The findings need to be supported by convincing arguments.

Definition 5.1 (Authorship attribution). Let $\{S_1, \cdots, S_n\}$ be the set of suspected authors of a malicious e-mail message ω. The assumption is that investigators have access to sample messages M_i, for each suspect $S_i \in \{S_1, \cdots, S_n\}$. The goal of authorship attribution is to identify the most plausible author S_a, from the suspects $\{S_1, \cdots, S_n\}$, in order to determine whose collection of messages M_a is the "best match" with the patterns in the malicious message ω. Intuitively, a collection of messages M_i matches ω if M_i and ω share similar patterns of stylometric features such as vocabulary usage.

The problem of authorship attribution can be refined into three subproblems: (1) Identifying the Writeprint $WP(M_i)$ from each set of e-mail messages $M_i \in \{M_1, \cdots, M_m\}$; (2) Determining the author of the malicious e-mail ω by matching ω with each of $\{WP(M_1), \cdots, WP(M_m)\}$; (3) Extracting evidence for supporting the conclusion on authorship. The evidence has to be intuitive enough to convince the judge and the jury in the court of law in order for them to make a judgment based on the facts of the case. These three subproblems summarize the challenges in a typical investigation procedure. To solve subproblems (1) and (2), the set of frequent patterns $FP(M_i)$ from M_i must first be extracted and then the patterns appearing in any other sets of e-mails M_j are filtered out. For subproblem (3), the Writeprint $WP(M_a)$ could serve as evidence for supporting the conclusion, where M_a is the set of e-mail messages written by the identified author S_a.

5.1.2   Attribution with Stylistic Variation

Although existing authorship studies mention the changing style of an author, they do not provide any conclusive solutions. In the current chapter, the problem of authorship attribution is defined with a focus on the problem of stylistic variation or volatile stylistics. The goal is to isolate the different sub-styles of a suspect, capture those styles, and then compute the Writeprint for each sub-style, called the sub-Writeprint of the suspect. Next, the anonymous message is compared with each sub-Writeprint to identify its true author. A more explicit description of the problem definition is given as follows.

Definition 5.2 (Authorship identification with stylistic variation). Suppose $\{S_1, \cdots, S_n\}$ is a set of suspected authors of an anonymous text message ω. Let $\{M_1, \cdots, M_n\}$ be the sets of text messages previously written by suspects $\{S_1, \cdots, S_n\}$, respectively. Let's assume that the message samples reflect the phenomenon of stylistic variation, which means that the collected messages contain different topics and are written to different types of recipients, e.g., co-workers and friends. In addition, let's assume that the number of messages within each set M_i, denoted by $|M_i|$, is reasonably large (say >30). The first task is to divide messages M_i from each suspect $S_i \in \{S_1, \cdots, S_n\}$ into k different groups $\{G_i^1, \cdots, G_i^k\}$ and then apply the attribution problem to identify the most plausible suspect $S_a \in \{S_1, \cdots, S_n\}$. The suspected author S_a is the one who's sub-Writeprint is the "best match" with the features in ω.

5.2   Building Blocks of the Proposed Approach

The core concepts of the proposed approach are feature extraction and feature discretization. The extracted features are used to identify frequent stylometric patterns and convert them into the Writeprint. A detailed description of these concepts is given below.

Feature Extraction   Feature extraction starts with message extraction, followed by cleaning, tokenization, and stemming, as discussed in sections below. There have been more than one thousand stylometric features used so far in different studies [63, 110].

As listed in Tables 5.1 and 5.2, 285 features are selected carefully. In general, there are three types of features. The first type is a numerical value, e.g., the frequencies of some individual characters, punctuations, and special characters. To avoid the situation where very large values, outweigh other features, normalization has been applied to scale down all the numerical values to [0, 1]. The second type is a Boolean value, e.g., to check whether an e-mail contains a reply message.

Table 5.1 Lexical and syntactic features

Feature type	Feature name
Lexical	1. Character count including space characters (M)
	2. Ratio of letters to M
	3. Ratio of uppercase letters to M
	4. Ratio of digits to M
	5. Ratio of tabs to M
	6. Ratio of spaces to M
	7–32. Alphabet frequency (A-Z) (26 features)
	33–53. Occurrences of special characters: < > % | { } [ ] / \ @ # ~ + - * $ ^ & _ ÷ (21 features)
	54. Word count (W)
	55. Average word length
	56. Average sentence-length in terms of characters
	57. Ratio of short words (1–3 characters) to W
	58–87. Ratio of word length frequency distribution to W (30 features)
	88. Ratio of function words to W
	89. Ratio of Hapax legomena to W
	90. Ratio of Hapax legomena to T
	91. Ratio of Hapax dislegomena to W
	92. Vocabulary richness, i.e., T/W
	93. Guirad's R
	94. Herdan's C
	95. Herdan's V
	96. Rubet's K
	97. Maas' A
	98. Dugast's U
Syntactic	99–106. Occurrences of punctuations, . ? ! : ; ' " (8 features)
	107. Ratio of punctuations with M
	108–257. Occurrences of function words (150 features)

The third type of feature is computed by taking as input some other lexical functions such as vocabulary richness, indexed at 93–98 in Table 5.1. Most of these features are computed in terms of vocabulary size V(N) and text length N [141]. When feature extraction is done, each e-mail is represented as a vector of feature values. This chapter focuses on using structural features as they play a significant role in distinguishing writing styles. Short words comprising of 1–3 characters (such as 'is', 'are', 'or', 'and', etc.) are mostly context-independent and are counted together. Frequencies of words of various lengths between 1–30 characters, indexed at 58–87 in Table 5.1, are counted separately. For once-occurring and twice-occurring words, the terms Hepax Legomena and Hapax dislegomena respectively are used. The welcoming and/or farewell greetings words in the e-mail are also considered.

A paragraph separator can be a blank line, tab/indentation or it is also possible for there to be no separator between paragraphs. Thirteen content-specific terms (273–285) are selected from the Enron e-mail corpus by applying content-based clustering. Each message is represented as a feature vector using the vector space model, as shown in Table 5.3. This table represents ten sample messages, where each row represents one e-mail message.

Table 5.2 Structural and domain-specific features

Feature type	Feature name
Structural	258. Ratio of blank lines/total number of lines within the e-mail
	259. Presence/absence of greetings
	260. Has tab as separators between paragraphs
	261. Has blank line between paragraphs
	262. Presence/absence of separator between paragraphs
	263. Average paragraph length in terms of characters
	264. Average paragraph length in terms of words
	265. Average paragraph length in terms of sentences
	266. Sentence count
	267. Paragraph count
	268. Contains Replied message
	269. The position of the replied message in the e-mail
	270. Use e-mail as a signature
	271. Use telephone as a signature
	272. Use URL as a signature
Domain-specific	273–285. the deal, HP, sale, payment, check, windows, software, offer, Microsoft, meeting, conference, room, report (13 features)

Table 5.3 Stylometric feature vectors (prior to discretization)

Messages (μ)	Feature X	Feature Y	Feature Z
μ_1	0.130	0.580	0.555
μ_2	0.132	0.010	0.001
μ_3	0.133	0.0124	0.123
μ_4	0.119	0.250	0.345
μ_5	0	0.236	0.532
μ_6	0.150	0.570	0.679
μ_7	0	0.022	0.673
μ_8	0.865	0.883	0.990
μ_9	0.137	0.444	0.494
μ_{10}	0.0	0.455	1.000

One may first apply feature selection as a preprocessing step to determine a subset of stylometric features that can discriminate the authors. There are two general approaches: Forward selection starts with no features and, at each step, it adds the feature that does best at reducing error until there is no significant decrease in error as a result of any further addition. Backward selection starts with all the features and, at each step, removes the one that reduces the error the most until any further removal causes error to increase significantly.

These approaches consider only one attribute at a time. In contrast, the proposed approach employs the notion of frequent stylometric patterns that capture the combined effect of features. Irrelevant features will not be frequent in this approach. Thus, there is no need to apply feature selection. More importantly, feature selection does not guarantee the property of uniqueness among the Writeprints of the suspects.

Feature Discretization The feature vectors, extracted in the previous step, contain numeric values. To extract frequent patterns from the message dataset, the Apriori algorithm has been applied [110]. For this, the numeric feature values need to be transformed into Boolean type values indicating the presence or absence of a feature within a message. Each feature $F_a \in \{F_1, \cdots, F_g\}$ is discretized into a set of intervals $\{\iota_1, \cdots, \iota_h\}$, called feature items. Common discretization techniques are:

- *Equal-width discretization*, where the size of each interval is the same.
- *Equal-frequency discretization*, where each interval has approximately the same number of records assigned to it.
- *Clustering-based discretization*, where clustering is performed on the distance of neighboring points.

Due to the small size of an e-mail message, most feature values fall into the beginning of an interval and need to be discretized in a more dynamic way. The initial experimental results indicate that the value of most features is close to zero with very few features having larger values. Therefore, employing equal-width discretization, and/or equal-frequency discretization is not the ideal choice while the clustering-based discretization method is complex and computationally expensive. To fit the niche, a new discretization mechanism called controlled binary split has been developed. This mechanism has substantially improved results compared to the initial study [64].

In the proposed technique, the feature value is successfully split into two intervals and investigators check to see if the number of feature occurrences is less than the user-specified threshold or not. Binary splitting continues until all feature values are discretized. The normalized feature frequency, found in a message, is then matched with these intervals. A Boolean '1' is assigned to the feature item if the interval contains the normalized feature frequency; otherwise, a '0' is assigned.

Example 5.1 Consider Table 5.4, which contains 10 e-mail messages. Let us assume that $\{X, Y, Z\}$ represent the set of features extracted from these messages. Next, each feature is converted into feature items by applying discretization. For example, feature X has normalized values in the range $[0, 1]$. Suppose the user threshold is 5%, i.e., the splitting continues until each interval contains at most 5% of the total number of feature occurrences. Feature X is discretized into three intervals $X_1 = [0, 0.120]$, $X_2 = (0.120, 0.140]$, and $X_3 = (0.140, 1.000]$, representing three feature items. Similarly, features Y and Z are discretized into $Y_1 = [0, 0.500]$, $Y_2 = (0.500, 1]$, $Z_1 = [0, 0.500]$, and $Z_2 = (0.500, 1]$, respectively. The message μ_1 containing features $X = 0.130$, $Y = 0.250$, and $Z = 0.020$ can be represented as a feature vector $\langle X_2, Y_1, Z_1 \rangle$.

Frequent Stylometric Patterns Intuitively, the stylometric patterns or the writing style patterns in an ensemble of e-mail messages M_i (written by suspect S_i) is a combination of feature items that frequently occurs in M_i. Such frequent patterns are concisely modeled and captured by the concept of frequent item set [110] described as follows. Let $U = \{\iota_1, \cdots, \iota_u\}$ denotes the universe of all stylometric feature items.

Table 5.4 Stylometric feature vectors (after discretization)

Messages (μ)	Feature X			Feature Y		Feature Z	
	X_1	X_2	X_3	Y_1	Y_2	Z_1	Z_2
μ_1	0	1	0	0	1	0	1
μ_2	0	1	0	1	0	1	0
μ_3	0	1	0	1	0	1	0
μ_4	1	0	0	1	0	1	0
μ_5	0	0	0	1	0	0	1
μ_6	0	0	1	0	1	0	1
μ_7	0	0	0	1	0	0	1
μ_8	0	0	1	0	1	0	1
μ_9	0	1	0	1	0	1	0
μ_{10}	0	0	0	1	0	0	1

Table 5.5 Message representation in terms of feature items

Messages (μ)	Features items
μ_1	$\{X_2, Y_2, Z_2\}$
μ_2	$\{X_2, Y_1, Z_1\}$
μ_3	$\{X_2, Y_1, Z_1\}$
μ_4	$\{X_1, Y_1, Z_1\}$
μ_5	$\{Y_1, Z_2\}$
μ_6	$\{X_3, Y_2, Z_2\}$
μ_7	$\{Y_1, Z_2\}$
μ_8	$\{X_3, Y_2, Z_2\}$
μ_9	$\{X_2, Y_1, Z_1\}$
μ_{10}	$\{X_1, Y_1, Z_2\}$

Let M_i be a set of e-mail messages where each message $\mu \in M_i$ is represented as a set of stylometric feature items such that $\mu \subseteq U$.

A text message μ contains a stylometric feature item ι_j if the numerical feature value of the message μ falls within the interval of ι_j. The writing style features of some sample messages are represented as vectors of feature items. The style features of some sample messages are represented as vectors of feature items in Table 5.5. Let $P \subseteq U$ be a set of stylometric feature items called a stylometric pattern. A text message μ contains a stylometric pattern P if $P \subseteq \mu$. A stylometric pattern that contains κ stylometric feature items is a κ-pattern. For example, the stylometric pattern $P = \{\iota_1, \iota_4, \iota_6\}$ is a three-pattern. The support of a stylometric pattern P is the percentage of text messages in M_i that contains P. A stylometric pattern P is frequent in a set of messages M_i if the support of P is greater than or equal to a user-specified minimum support threshold.

Definition 5.3 (Frequent stylometric pattern). Let M_i be the set of text messages written by suspect S_i. Let support ($P|M_i$) be the percentage of text messages in M_i that contain the pattern P, where $P \subseteq U$. A pattern P is a frequent stylometric pattern in M_i if support ($P|M_i$) $\geq$ min_sup, where the minimum support threshold min_sup is a real number in an interval of [0, 1]. The writing style of a suspect S_i is represented as a set of frequent stylometric patterns, denoted by $FP(M_i) = \{P_1, \cdots, P_l\}$, extracted from his/her set of text messages.

Example 5.2 Consider the messages, represented as vectors of feature items, in Table 5.5. Suppose the user-specified threshold min_sup = 0.3, which means that a stylometric pattern $P = \{\iota_1, \cdots, \iota_e\}$ is frequent if at least 3 out of the 10 e-mails contain all feature items in P. For instance, $\{X_1\}$ is not a frequent stylometric pattern because it has support 2/10 = 0.2. The feature item $\{X_2\}$ is a frequent stylometric one-pattern because it has a support value of 0.4. Similarly, $\{X_2, Y_1\}$ is a frequent stylometric two-pattern because it has support 0.4. Likewise, $\{X_2, Y_1, Z_1\}$ is a frequent stylometric three-pattern because it has a support value of 0.3. Example 5.1 shows how to efficiently compute all frequent patterns.

5.3   Writeprint

In forensic science, the fingerprints of a person can uniquely identify him/her. In cyber forensics, the "Writeprint" of a person can be identified using his/her e-mails. We do not claim that the identified Writeprints in this study are able to uniquely identify every individual in the world, but they are precise enough to uniquely identify the writing pattern of an individual among the suspects $\{S_1, \cdots, S_n\}$ because common patterns among the suspects are filtered out and will not become part of the Writeprint.

The notion of frequent pattern in Definition 5.1 captures the writing patterns of a suspect. However, two suspects S_i and S_j may share some similar writing patterns. Therefore, it is important to filter out the common frequent patterns and retain the frequent patterns that are unique to each suspect. This leads us to the notion of Writeprint.

Intuitively, a Writeprint can uniquely represent the writing style of a suspect S_i if its patterns are found only in the e-mails written by S_i, but not in any other suspect's e-mails. In other words, the Writeprint of a suspect S_i is a collection of frequent patterns that are frequent in the e-mail messages M_i written by S_i but not frequent in the messages M_j written by any other suspect S_j where $i \neq j$.

Definition 5.4 (Writeprint). A Writeprint, denoted by $WP(M_i)$, is a set of patterns where each pattern P has support$(P|M_i) \geq$ min_sup and support$(P|M_j) <$ min_sup for any M_j where $i \neq j$, min_sup is a user-specified minimum threshold. In other words, $WP(M_i) \subseteq FP(M_i)$, and $WP(M_i) \cap WP(M_j) = \emptyset$ for any $1 \leq i, j \leq n$ and $i \neq j$.

5.4   Proposed Approaches

The proposed solution is divided into two parts. The first part, called AuthorMiner1, addresses the traditional attribution problem without the consideration of stylistic variation while the second part, called AuthorMiner2, addresses the attribution problem with the consideration of stylistic variation. A detailed description of the two components is given in the following two subsections.

5.4.1 AuthorMiner1: Attribution Without Stylistic Variation

Algorithm 5.1 presents a novel machine learning method, called AuthorMiner1, for
determining the authorship of a malicious e-mail message ω from a group of suspects $\{S_1, \cdots, S_n\}$ based on the extracted features of their previously written e-mail
messages $\{M_1, \cdots, M_n\}$. In this section, an e-mail message is represented by a set of
featured items. The algorithm has been summarized in the following three phases
followed by a detailed description of each phase.

Phase 1: Mining frequent patterns (Lines 1–3) Extract the frequent patterns
$FP(M_i)$ from each collection of e-mail messages M_i written by suspect S_i. The
extracted frequent patterns capture the writing style of a suspect.

Algorithm 5.1 AuthorMiner1

Require: An anonymous message ω.
Require: Sets of messages $\{M_1, \cdots, M_n\}$, written by $\{S_1, \cdots, S_n\}$.
 /* Mining frequent stylometric patterns */
 1: **for each** $M_i \in \{M_1, \cdots, M_n\}$ **do**
 2:     extract frequent stylometric patterns $FP(M_i)$ from M_i;
 3: **end for**
 /* Filtering out common frequent patterns */
 4:   **for each** $FP(M_i) \in \{FP(M_1), \cdots, FP(M_n)\}$ **do**
 5:     **for each** $FP(M_j) \in \{FP(M_{i+1}), \cdots, FP(M_n)\}$ **do**
 6:         **for each** frequent pattern $P_x \in FP(M_i)$ **do**
 7:             **for each** frequent pattern $P_y \in FP(M_j)$ **do**
 8:                 **if** $P_x = P_y$ **then**
 9:                     $FP(M_i) \leftarrow FP(M_i) - P_x$;
10:                     $FP(M_j) \leftarrow FP(M_j) - P_y$;
11:                 **end if**
12:             **end for**
13:         **end for**
14:     **end for**
15:     $WP(M_i) \leftarrow$ *Disjoint frequent patterns(M_i)*;
16: **end for**
 /* Identifying author */
17: *highest_score* $\leftarrow -1$;
18: **for all** $WP(M_i) \in \{WP(M_1), \cdots, WP(M_n)\}$ **do**   19:   **if** $Score(ω \approx WP(M_i)) > highest\_score$ **then**
20:     *highest_score* $\leftarrow Score(ω \approx WP(M_i))$;
21:         *author* $\leftarrow S_i$;
22:     **end if**
23: **end for**
24: **return** *author*;

Phase 2: Filtering common frequent patterns (Lines 4–16) Though $FP(M_i)$ may capture the writing patterns of suspect S_i, $FP(M_i)$ may contain frequent patterns that are shared by other suspects. Therefore, Phase 2 removes the common frequent patterns. Specifically, a pattern P in $FP(M_i)$ is removed if any other $FP(M_j)$ also contains P, where $i \neq j$. The remaining frequent patterns in $FP(M_i)$ form the Writeprint $WP(M_i)$ of suspect S_i. When this phase is complete, there is a set of Writeprints $\{WP(M_1), \cdots, WP(M_n)\}$ of suspects $\{S_1, \cdots, S_n\}$. Figure 5.1 illustrates that the Writeprint $WP(M_2)$ comes from $FP(M_2)$ by filtering out the frequent patterns shared by $FP(M_1)$, $FP(M_2)$, and/or $FP(M_3)$.

Phase 3: Identifying the author (Lines 17–24) Compare the malicious e-mail message ω with each Writeprint $WP(M_i) \in \{WP(M_1), \cdots, WP(M_n)\}$ and identify the most similar Writeprint that matches ω. Intuitively, a Writeprint $WP(M_i)$ is similar to the e-mail message ω if many frequent patterns in $WP(M_i)$ can be found in ω.

It is important to note that frequent patterns are not equally important. Their importance is reflected by their support($P|M_i$); therefore, a score function Score($\omega \approx WP(M_i)$) is derived to measure the weighted similarity between the e-mail message ω and the frequent patterns in $WP(M_i)$. The suspect S_a of Writeprint $WP(M_a)$, which has the highest Score($\omega \approx WP(M_i)$), is deemed to be the author of the malicious e-mail message ω.

Mining Frequent Stylometric Patterns (Lines 1–3) Lines 1–3 mine the frequent patterns $FP(M_i)$ from each collection of e-mail message $M_i \in \{M_1, \cdots, M_n\}$, for $1 \leq i \leq n$.

There are many machine learning and data mining algorithms for extracting frequent patterns, for example, Apriori [139], FP-Growth [142], and ECLAT [143]. Below, an overview of the Apriori algorithm is provided which has been previously applied to various text mining tasks [144, 145]. Apriori is a level-wise iterative search algorithm that uses frequent κ-patterns to explore the frequent ($\kappa + 1$)-patterns. First, the set of frequent one-patterns is found by scanning the e-mail messages M_i, accumulating the support count of each feature item, and collecting the feature item ι that has support($\iota|M_i$) $\geq$ min_sup. The resulting frequent one-patterns are then used to find frequent two-patterns, which are then used to find frequent three-patterns, and so on until no more frequent κ-patterns can be found. The generation of frequent ($\kappa + 1$)-patterns from frequent κ-patterns is based on the following Apriori property.

Property 5.1 (Apriori property). All nonempty subsets of a frequent pattern must also be frequent by definition. A pattern P is not frequent if support ($P|M_i$) < min_sup. The above property implies that adding a feature item ι to a non-frequent pattern P will never make it more frequent. Thus, if a κ-pattern P is not frequent, then there is no need to generate ($\kappa + 1$)-pattern $P \cup \iota$ because $P \cup \iota$ is also not frequent. The following example shows how the Apriori algorithm exploits this property to efficiently extract all frequent patterns. Refer to [139] for a formal description.

Example 5.3 Consider Table 5.5 with min_sup = 0.3. First, identify all frequent one-patterns by scanning the database once to obtain the support of every feature item. The feature items having support ≥ 0.3 are frequent one-patterns, denoted by $L_1 = \{\{X_2\}, \{Y_1\}, \{Z_1\}, \{Z_2\}\}$. Then, join L_1 with itself, i.e., $L_1 \bowtie L_1$, to generate the candidate list $L_2 = \{\{X_2, Y_1\}, \{X_2, Z_1\}, \{X_2, Z_2\}, \{Y_1, Z_1\}, \{Y_1, Z_2\}, \{Z_1, Z_2\}\}$ and scan the database once to obtain the support of every pattern in L_2. Identify the frequent two-patterns, denoted by $L_2 = \{\{X_2, Y_1\}, \{X_2, Z_1\}, \{Y_1, Z_1\}, \{Y_1, Z_2\}\}$. Similarly, perform $L_2 \bowtie L_2$ to generate ℓ_3 and scan the database once to identify the frequent three-patterns which is $L_3 = \{X_2, Y_1, Z_1\}$. The finding of each set of frequent κ-patterns requires one full scan of the feature items in Table 5.5.

Filtering Common Patterns (Lines 4–16) This phase filters out the common frequent patterns among $\{FP(M_1), \cdots, FP(M_n)\}$. The general idea is to compare every frequent pattern P_x in $FP(M_i)$ with every frequent pattern P_y in all other $FP(M_j)$ and to remove them from $FP(M_i)$ and $FP(M_j)$ if P_x and P_y are the same. The computational complexity of this step is $O(|FP(M)|^n)$, where $|FP(M)|$ is the number of frequent patterns in $FP(M)$ and n is the number of suspects. The remaining frequent patterns in $FP(M_i)$ form the Writeprint $WP(M_i)$ of suspect S_i.

Example 5.4 Suppose there are three suspects S_1, S_2, and S_3 with three sets of e-mail messages M_1, M_2, and M_3 respectively, as depicted in Fig. 5.1. Let $FP(M_1) = \{\{X_1\}, \{Y_1\}, \{Z_2\}, \{X_1, Y_1\}, \{X_1, Z_2\}, \{Y_1, Z_2\}, \{X_1, Y_1, Z_2\}\}$ be the frequent patterns of S_1. Let $FP(M_2) = \{\{X_2\}, \{Y_1\}, \{Z_1\}, \{Z_2\}, \{X_2, Y_1\}, \{X_2, Z_1\}, \{Y_1, Z_1\}, \{Y_1, Z_2\}, \{X_2, Y_1, Z_1\}\}$ be the set of frequent patterns of S_2, as given in Example 5.2. Let $FP(M_3) = \{\{X_1\}, \{Y_3\}, \{Z_2\}, \{X_1, Y_3\}, \{X_1, Z_2\}, \{Y_3, Z_2\}, \{X_1, Y_3, Z_2\}\}$ be the set of frequent patterns of S_3. Then, $\{X_1\}, \{Y_1\}, \{Z_2\}, \{X_1, Z_2\}, \{Y_1, Z_2\}$ are discarded as they are shared by two or more suspects. The remaining frequent patterns form the Writeprints of the suspects: $WP(M_1) = \{\{X_1, Y_1\}, \{X_1, Y_1, Z_2\}\}$, $WP(M_2) = \{\{X_2\}, \{Z_1\}, \{X_2, Y_1\}, \{X_2, Z_1\}, \{Y_1, Z_1\}, \{X_2, Y_1, Z_1\}\}$, and $WP(M_3) = \{\{Y_3\}, \{X_1, Y_3\}, \{Y_3, Z_2\}, \{X_1, Y_3, Z_2\}\}$.

Identifying Author (Lines 17–24) Lines 17–24 determine the author of malicious e-mail message ω by comparing ω with each Writeprint $WP(M_i) \in \{WP(M_1), \cdots, WP(M_n)\}$ and identifying the most similar Writeprint to ω. Intuitively, a Writeprint $WP(M_i)$ is similar to ω if many frequent patterns in $WP(M_i)$ matches the style in ω. Formally, a frequent pattern P matches ω if ω contains every featured item in P. Equation 5.1 shows the score function that quantifies the similarity between the malicious message ω and a Writeprint $WP(M_i)$. The frequent patterns are not equally important, and their importance is reflected by their support in M_i, i.e., the percentage of e-mail messages in M_i sharing such a combination of features. Thus, the score function accumulates the support of a frequent pattern and divides the result by the number of frequent patterns in $WP(M_i)$ to normalize the factor of different sized $WP(M_i)$.

$$Score\left(\omega \approx WP\left(M_i\right)\right) = \frac{\sum_{j=1}^{\rho} support(MP_j | M_i)}{\left|WP\left(M_i\right)\right|} \qquad (5.1)$$

where $MP = \{MP_1, \cdots, MP_\rho\}$ is a set of matched patterns between $WP(M_i)$ and the malicious e-mail message ω. A score is a real number within the range $[0, 1]$. The higher the score means the higher the similarity between the Writeprint and the malicious e-mail message ω. The suspect having the Writeprint with the highest score is the author of the malicious e-mail ω.

Example 5.5 Let the patterns found in the malicious e-mail message ω be $\{X_2, Y_1, Z_1\}$ and $\{X_1, Y_1, Z_2\}$. Comparing them to the Writeprints in Example 5.4, it has been noticed that the first pattern matches with a pattern in $WP(M_2)$ while the second pattern matches with a pattern in $WP(M_1)$. The score calculated according to Eq. 5.1 is higher for $WP(M_1)$ because $|WP(M_1)| < |WP(M_2)|$. As a result, the message ω is the most similar to $WP(M_1)$, suggesting that S_1 is its author.

In the unlikely case that multiple suspects have the same highest score, AuthorMiner1 returns the suspect whose number of matched patterns $|MP|$ is the largest. In case multiple suspects have the same highest score and the same number of matched patterns, AuthorMiner1 returns the suspect whose size of matched k-pattern is the largest because having a match on large-sized frequent stylometric k-pattern implies a strong match. To facilitate the evaluation procedure in this experiment, the method presented here is designed to return only one suspect. In the actual deployment of the method, a preferable solution is to return a list of suspects ranked by their scores, followed by the number of matched patterns and the size of the largest matched pattern.

5.4.2 AuthorMiner2: Attribution with Stylistic Variation

In AuthorMiner2, the focus is on the occasional change in the writing style of individuals due to the change in the context and/or target recipient. The change may occur both in the content as well as in the style markers. For instance, e-mails that a person writes to his colleagues are more formal than e-mails that he writes to his family members and friends. Co-workers of a financial company may write more about meetings, promotion schemes, customer problems and solutions, salaries, and bonuses. E-mails exchanged among friends may contain discussions about trips, visits, funny stories, and jokes.

Writing style features such as the selection and distribution of function words and punctuation may be different in different contexts. Moreover, an employee may be more formal and careful in using structural features such as greeting and farewell comments in e-mails written to upper management, for example. An employee may prefer to include complete signatures including job title and contact information in work-related communication. More importantly, malicious e-mails are mostly anonymous and will often be devoid of such traceable information.

In fact, information (topic words and stylometric features) extracted from malicious messages is overshadowed by information in regular messages, as malicious messages are usually much fewer in number than regular messages. If the same analytical techniques were used for such different types of writing samples, misleading results might be produced.

To address the authorship problem, the algorithm has been proposed, called *AuthorMiner2*. *AuthorMiner2* is employed to capture the different writing styles, called the sub-styles, of a person, and thus the authorship identification accuracy can be improved. The experimental results generated support the hypothesis and suggest that the author identification accuracy of *AuthorMiner2* is higher than *AuthorMiner1*. This is because *AuthorMiner2* can be employed to concisely present the fine-grained writing styles of an individual.

Grouping based on Message Body In *AuthorMiner2*, two types of clustering are applied: content-based and stylometry-based. In content-based clustering, the messages are divided into different groups based on the topic of discussion [115]. Stylometry-based clustering, on the other hand, is used to divide messages into different groups; each group containing similar patterns of writing style features [57].

The process of clustering in both cases is the same. The difference is in the details of the preprocessing and feature extraction phase. In content-based clustering, the preprocessing step is similar to the usual text mining process where the style markers (function words and punctuations) and white and blank spaces are deleted along with other irrelevant parts of a document. The remaining content is tokenized and stemmed to obtain a list of topic words. The preprocessing phase in stylometry-based clustering is complex where most of the message content, including topic words and style markers, are used as features. Once all the e-mail messages of each author are converted into feature vectors, clustering is applied. Three clustering algorithms are used: Expectation Maximization (EM), k-means, and bisecting k-means. Clustering is applied to the e-mails of each author independently. The resultant clusters of each suspect S_i are labeled as $\{G_i^1, \cdots, G_i^k\}$. Similarly, e-mail messages of S_j are clustered separately into clusters $\{G_j^1, \cdots, G_j^k\}$.

Grouping based on Message Header The e-mail messages of each suspect are divided into different groups based on header-content including *e-mail recipient* and *e-mail timestamp*. The reasoning behind using the timestamp for grouping is that some researchers, like J. Stolfo et al. [146] believe that people behave differently at different times of the day and night.

People usually communicate with different categories of people at different times. For instance, most of the e-mails that a person writes during the daytime are exchanged with co-workers. Similarly, e-mail messages written in the evening may be exchanged with family members and friends. Likewise, in most cases, very few of the e-mail messages that are exchanged at midnight are written to colleagues. For simplicity, the 24 h are divided into three-time brackets: morning, evening, and night. Therefore, a sender's e-mails are divided into three categories: e-mails sent in the morning, e-mails sent in the evening, and those sent at night.

Extracting Frequent Stylometric Patterns Step 2 (Lines 3–6 in Algorithm 5.2) extracts the frequent stylometric patterns from each group $G_i{}^g$ for each message set M_i of suspect S_i. Frequent stylometric patterns $\{FP(G_i{}^1), \cdots, FP(G_i{}^k)\}$ from message subsets $\{G_i{}^1, \cdots, G_i{}^k\}$ of suspect S_i are extracted by using this technique.

Algorithm 5.2 AuthorMiner2

Input: An anonymous message ω
Input: Messages $\{M_1, \cdots, M_n\}$ by $\{S_1, \cdots, S_n\}$.

1: **for all** $M_i \in \{M_1, \cdots, M_n\}$ **do**
2:    Divide M_i into groups $\{G_i^1, \cdots, G_i^k\}$;
3:    **for all** $G_i^g \in \{G_i^1, \cdots, G_i^k\}$ **do**
4:        extract frequent stylometric patterns $FP(G_i^g)$ from G_i^g;
5:    **end for**
6: **end for**
7: **for all** $M_i \in \{M_1, \cdots, M_n\}$ **do**
8:    **for all** $G_i^g \in \{G_i^1, \cdots, G_i^k\}$ **do**
9:        **for all** $M_j \in \{M_{i+1}, \cdots, M_n\}$ **do**
10:            **for all** $G_j^h \in \{G_j^1, \cdots, G_j^k\}$ **do**
11:                **if** $G_i^g \neq G_j^h$ **then**
12:                    **for all** frequent stylometric pattern $P_x \in FP(G_i^g)$ **do**
13:                        **for all** frequent stylometric pattern $P_y \in FP(G_j^h)$ **do**
14:                            **if** $P_x = P_y$ **then**
15:                                $FP(G_i^g) \leftarrow FP(G_i^g) - P_x$;
16:                                $FP(G_j^h) \leftarrow FP(G_j^h) - P_y$;
17:                            **end if**
18:                        **end for**
19:                    **end for**
20:                **end if**
21:            **end for**
22:        **end for**
23:        $WP(G_i^g) \leftarrow Disjoint\ set\ of\ FP(G_i^g)$;
24:    **end for**
25: **end for**
26: $highest\_score \leftarrow -1$;
27: **for all** $M_i \in \{M_1, \cdots, M_n\}$ **do**
28:    **for all** $G_i^g \in \{G_i^1, \cdots, G_i^k\}$ **do**
29:        **if** $Score(\omega \approx WP(G_i^g)) > highest\_score$ **then**
30:            $highest\_score \leftarrow Score(\omega \approx WP(G_i^g))$;
31:            $author \leftarrow S_i$;
32:        **end if**
33:    **end for**
34: **end for**
35: **return** $author$;

Filtering Common Stylometric Patterns Step 3 (Lines 7–25 in Algorithm 5.2) filters out the common stylometric frequent patterns between any two sets $FP(G_i^g)$ and $FP(G_j^h)$ where $i \neq j$. As described in Sect. 5.1, the general idea is to compare every frequent pattern P_x in $FP(G_i^g)$ with every frequent pattern P_y in all other sets, e.g., $FP(G_j^h)$, and to remove them from $FP(G_i^g)$ and $FP(G_j^h)$ if P_x and P_y are the same. The computational complexity of this step is O ($|\cup FP(G_i^g)|^2$), where $|\cup FP(G_i^g)|$ is the total number of stylometric frequent patterns. The remaining stylometric frequent patterns in $FP(G_i^g)$ represents a sub-Writeprint $WP(G_i^g)$ of suspect S_i. A suspect S_i may have multiple sub-Writeprints, denoted by $\{WP(G_i^1), \cdots, WP(G_i^k)\}$ depending on how the messages are grouped in Step 1.

$$Score\left(\omega \approx WP\left(G_i^g\right)\right) = \frac{\sum_{j=1}^{\rho} support(MP_j | G_i^g)}{\left|WP\left(G_i^g\right)\right|} \qquad (5.2)$$

Identifying Author Step 4 (Lines 26–35 in Algorithm 5.2) determines the author of the anonymous message ω by comparing ω with each Writeprint $WP(G_i^g)$ of every suspect S_i and identifying the Writeprint that is similar to ω. Intuitively, a Writeprint $WP(G_i^g)$ is similar to ω if many frequent stylometric patterns in $WP(G_i^g)$ match the stylometric feature items found in ω. The score function in Eq. 5.2 is the modified form of Eq. 5.1, which is used to measure the similarity between the anonymous message ω and a Writeprint $WP(G_i^g)$. The proposed score function accumulates the support count of a frequent stylometric pattern.

To conclude this chapter, the proposed solution was divided into two parts. The first part, called AuthorMiner1, addressed the traditional attribution problem without the consideration of stylistic variation while the second part, called AuthorMiner2, addressed the attribution problem with the consideration of stylistic variation. This chapter defined two authorship identification problems: Firstly, the attribution of an anonymous message to the true author by ignoring the occasional stylistic variation of the potential authors. Secondly, the attribution of an anonymous message with a contextual stylistic change of potential suspects. The first problem was further refined into three sub-problems: (1) extracting the Writeprint of a suspect; (2) identifying the author of a malicious e-mail; and (3) collecting evidence for supporting the conclusion on authorship. Generally, the same methodology is applied in the court of law for resolving attribution issues. Most previous contributions focused on improving the classification accuracy of authorship identification, but only a few of them studied how to gather strong evidence for the court of law.

To address the first problem, the chapter introduced a novel approach of authorship attribution and formulated a new notion of Writeprint based on the concept of frequent patterns. To address the second problem, the chapter extended and improved the approach of frequent-pattern based Writeprint to capture the sub-styles of a suspect by creating the sub-Writeprint of a suspect.

Chapter 6
Authorship Attribution With Few Training Samples

This chapter discusses authorship attribution through a training sample. The focus on authorship attribution discussed in this chapter differs in two ways from the traditional authorship identification problem discussed in the earlier chapters of this book. Firstly, the traditional authorship attribution studies [63, 65] only work in the presence of large training samples from each candidate author, which are typically enough to build a classification model. With authorship attribution, the emphasis is on using a few training samples for each suspect. In some scenarios, no training samples may exist, and the suspects may be asked (usually through court orders) to produce a writing sample for investigation purposes. Secondly, in traditional authorship studies, the goal is to attribute a single anonymous document to its true author. In this chapter, we look at cases where we have more than one anonymous message that needs to be attributed to the true author(s). It is likely that the perpetrator may either create a ghost e-mail account or hack an existing account, and then use it for sending illegitimate messages in order to remain anonymous. To address the aforementioned shortfalls, the authorship attribution problem has been redefined as follows: given a collection of anonymous messages potentially written by a set of suspects $\{S_1, \cdots, S_n\}$, a cybercrime investigator first wants to identify the major groups of messages based on stylometric features; intuitively, each message group is written by one suspect. Then s/he wants to identify the author of each anonymous message collection from the given candidate suspects. To address the newly defined authorship attribution problem, the stylometric pattern-based approach of AuthorMinerl (described previously in Sect. 5.4.1) is extended and called AuthorMinerSmall. When applying this approach, the stylometric features are first extracted from the given anonymous message collection Ω.

As described earlier, stylometric features include lexical features, style markers (punctuation and function words), structural features, and content-specific features. Each message is converted into a feature vector using the vector space model representation. Then, stylometry-based clustering is applied to cluster the given messages

F. Iqbal et al., *Machine Learning for Authorship Attribution and Cyber Forensics*, International Series on Computer Entertainment and Media Technology, https://doi.org/10.1007/978-3-030-61675-5_6

into different groups. The idea is that clustering by stylometric features can group the messages by the same author together. The subsequent steps of the proposed method are applicable only if this hypothesis is true. The experimental results support the hypothesis.

Note that clustering applied in this chapter is different from traditional text clustering [144, 147] in two ways. Firstly, the objective of traditional clustering is to identify the different topics contained in the documents in question. The purpose of clustering in this context is to identify pertinent writing styles in the messages.

Secondly, traditional clustering is applied based on content-specific words, while in this case, clustering is applied based on stylometric features. Messages from each cluster are used to extract the frequent stylometric patterns by applying the first approach, AuthorMiner1, as described in the previous chapter. To compute the Writeprint of a cluster $C_i \in \{C_1, \cdots, C_k\}$, the stylometric patterns shared by more than one cluster are deleted. Next, the stylometric patterns $\{P(M_1), \cdots, P(M_k)\}$ from the training samples $\{M_1, \cdots, M_k\}$ of the suspects are extracted. Finally, each Writeprint $WP(C_j)$ is compared with every pattern $P(M_i) \in \{P(M_1), \cdots, P(M_k)\}$ to identify the most conceivable author S_a of cluster C_j.

Cluster analysis provides the crime investigator with a deep insight into the writing styles found in the given anonymous e-mails, in which the clusters and the extracted Writeprint could serve as input information for higher-level data mining. To investigate the relative discriminating power of different stylometric features, clustering is applied to each feature type (i.e., lexical, syntactic, structural, and content-specific) separately. In these experiments, the effects of the number of authors and the size of the training set on the purity of clusters are measured. Using the visualization and browsing features of the developed tool, an investigator can explore the process of cluster formation and evaluation. The contributions of this method are discussed as follows:

- **Attribution based on few training samples:** Existing authorship identification methods often require a comparatively large number of training samples in order to build a classification model. The proposed method is effective even if only a few training samples exist.
- **Clustering using stylometric features:** In the data mining community, content-based clustering is used to cluster messages into different groups based on the topic. Employing the same notion, the experimental results achieved using a real-life e-mail corpus (Enron e-mail corpus) [148] suggest that clustering by stylometric features is a sensible method of grouping together messages written by the same person.
- **Cluster analysis:** A method has been proposed and a tool has been developed for an investigator to visualize, browse, and explore the writing styles extracted from a collection of anonymous e-mails. The relative strength of different clustering algorithms is evaluated. This study reveals the relative discriminating power of four different categories of stylometric features. The effects of the number of suspects as well as the number of messages per suspect on the clustering accuracy are studied.

- **Dataset attribution:** The proposed method can be used to attribute a collection of messages (e.g., anonymous message or a ghost e-mail account) to its plausible author.

When an investigator has a disputed anonymous online message together with some potential suspects, his task is to identify the true author of the document in question by analyzing the sample documents of potential suspects. While existing authorship studies mention the temporal and contextual variation in the writing style of an author, they do not consider them in their solutions. This chapter discusses and addresses the problem of authorship attribution with and without a focus on the problem of stylistic variation. The term stylistic variation is used to represent the temporal and occasional change in the writing style of an individual. Punctuation and n-gram features have proven to be highly representative within their own context. The relative preference for using certain words over others along with their associations is another highly representative feature, which can help segregate authors' work.

As stated in the previous chapter, authorship attribution problem can be divided into four phases: message collection, feature extraction, model generation, and author identification. In most real-world investigation problems, the number of sample documents is often insufficient for training a classifier. In certain situations, the available sample may be very small or there may be no sample. In some cases, an investigator can ask a suspect to produce a sample of his or her writing by listening to a story or watching a movie and then reproducing the played scene in his/her own writing. Clearly, the number of samples is very limited.

6.1 Problem Statement and Fundamentals

This section gives a fundamental idea of existing authorship classification techniques. The goal of authorship attribution with few training samples is to identify the most plausible author S_a of a set of anonymous text messages Q from a group of suspects $\{S_1, \cdots, S_n\}$, with only a few sample text messages M_i for each suspect S_i. Note that this problem is different from the first problem defined above: (1) the number of training samples is small (say less than 30 sample e-mails). Therefore, it is infeasible to build a classifier as in the traditional classification method or to extract the frequent stylometric patterns based on low support counts. (2) The first problem focuses on how to identify the author of one anonymous message. In contrast, this problem focuses on how to cluster multiple anonymous text messages by stylometric features such that the messages of each cluster are written by the same author, and how to identify the author of each cluster of anonymous messages. The investigator needs to support his findings with convincing evidence. The problem is formally described as follows.

Definition 6.1 (Authorship attribution with few training samples). Let Ω be a set of anonymous text messages. Let $\{S_1, \cdots, S_n\}$ be a set of suspected authors of Ω. Let $\{M_1, \cdots, M_n\}$ be the sets of text messages previously written by suspects

$\{S_1, \cdots, S_n\}$, respectively. Assume $|M_i|$ is very small. The problem is to first group the messages Ω into clusters $\{C_1, \cdots, C_k\}$ by stylometric features, and then to identify the plausible author S_a from $\{S_1, \cdots, S_n\}$ for each cluster of anonymous messages $C_j \in \{C_1, \cdots, C_k\}$, with presentable evidence. The most plausible author S_a of C_j is the suspect whose stylometric patterns $P(M_i)$ have the "best match" with Writeprint $WP(C_j)$. To understand the problem with authorship attribution or authorship identification in the context of online documents refer to Chap. 5.

6.2 Proposed Approach

The general idea of this proposed method, depicted in Fig. 6.1, is composed of five steps. Step 1 involves preprocessing, feature extraction, and normalization. Step 2 is grouping anonymous messages Ω into clusters $\{C_1, \cdots, C_k\}$ by stylomet-

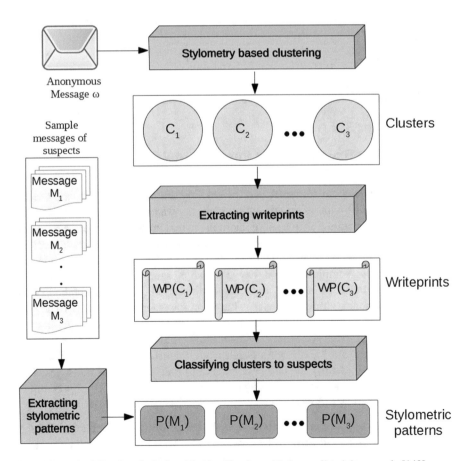

Fig. 6.1 AuthorMinerSmall: Authorship identification with the small training sample [149]

ric features such that each cluster contains the anonymous messages written by the same suspect. Step 3 is feature discretization and frequent stylometric patterns mining from each cluster of messages. Step 4 is calculating the Writeprint of each cluster by filtering the frequent stylometric patterns shared by two or more clusters. Step 5 is identifying the most plausible author Sa of each cluster C_j by comparing the extracted Writeprint $WP(C_j)$ with every set of training samples $M_i \in \{M_1, \cdots, M_n\}$.

6.2.1 Preprocessing

The preprocessing applied in this section is different from the preprocessing applied in the previous chapter. In this chapter, discretization has not been applied after the usual process of cleaning, tokenization, stemming, and feature extraction. Discretization is applied after the clusters are formed in the next section. Similarly, the preprocessing step of stylometry-based clustering [149] is different from traditional text clustering [115]. In traditional text clustering, only the content-specific words are counted while in stylometry-based clustering, the stylometric features are extracted in addition to the content-specific words. Using vector space model representation, each message μ is converted into a 285-dimensional vector of features $\mu = \{X_i, Y_j, Z_k\}$, as shown in Table 6.2. When all messages are converted into feature vectors, normalization is applied to the columns as needed. Discretization of the extracted features $\{X,Y,Z\}$ into respective feature items is done after the clustering phase.

Table 6.1 Clusters with member messages

Cluster C	Message (μ)	Feature values
C_1	μ_1	{0.130,0.580,0.555}
C_1	μ_2	{0.132,0.010,0.001}
C_1	μ_3	{0.133,0.0124,0.123}
C_2	μ_4	{0.119,0.250,0.345}
C_2	μ_5	{0.0,0.236,0.532}
C_2	μ_6	{0.150,0.570,0.679}
C_3	μ_7	{0.0,0.022,0.673}
C_3	μ_8	{0.985,0.883,0.990}
C_3	μ_9	{0.137,0.444,0.894}
C_3	μ_{10}	{0.0,0.455, 1.000}
C_3	μ_{11}	{0.134,0.012,0.0}
C_3	μ_{12}	{0.0,0.123, 1.000}

Table 6.2 Clustered messages after discretization

Cluster C	Message (μ)	Stylometric features
C_1	μ_1	$\{X_2, Y_2, Z_2\}$
C_1	μ_2	$\{X_2, Y_1, Z_1\}$
C_1	μ_3	$\{X_2, Y_1, Z_1\}$
C_1	μ_4	$\{X_1, Y_1, Z_1\}$
C_2	μ_5	$\{Y_1, Z_2\}$
C_2	μ_6	$\{X_3, Y_2, Z_2\}$
C_2	μ_7	$\{Y_1, Z_2\}$
C_2	μ_8	$\{X_3, Y_2, Z_2\}$
C_3	μ_9	$\{X_2, Y_1, Z_3\}$
C_3	μ_{10}	$\{Y_1, Z_3\}$
C_3	μ_{11}	$\{X_2, Y_1\}$
C_3	μ_{12}	$\{Y_1, Z_3\}$

6.2.2 *Clustering by Stylometric Features*

Statistic clustering groups the anonymous messages Ω into different clusters $\{C_1, \cdots, C_k\}$ on the basis of stylometric features. The hypothesis is that the writing style of every suspect is different, so clustering by stylometric features could group the messages written by the same author into one cluster. This clustering step is very different from AuthorMiner1 which groups training samples with the goal of identifying the sub-Writeprints of a suspect. In contrast, the reason for clustering anonymous messages in AuthorMinerSmall is to facilitate more precise Writeprint extraction, which is otherwise impossible due to the small set of training data. One can apply any clustering methods, such as k-means, to group the anonymous messages into clusters $\{C_1, \cdots, C_k\}$ such that messages in the same cluster have similar stylometric features and messages in different clusters have different stylometric features. Often, k is an input parameter to a clustering algorithm. In this case, k can be the number of suspects.

The proposed method is evaluated by employing three clustering algorithms: Expectation Maximization (EM), k-means, and bisecting *k-means*. The k-means clustering algorithm has been chosen [150] because it is known to be both simple and effective. The *k*means algorithm partitions a set of objects into *k* sub-classes. It attempts to find the centers of natural clusters in the data by assuming that the object attributes form a vector space and minimizing the intra-cluster variance. Thus, with the k-means algorithm, circular clusters tend to form around a centroid, and the algorithm outputs the centroids. k-means is particularly applicable to numeric attributes. The Expectation Maximization (EM) algorithm, first proposed in [151], is often employed where it is hard to predict the value of k (number of clusters). For instance, during the forensic analysis of anonymous e-mails, an investigator may not know the total number of suspects (or different writing styles) within a collection. In a more common scenario, a user may want to validate the results obtained by other clustering algorithms, say k-means or bisecting k-means. Suppose there are

12 anonymous messages and, after applying clustering, three clusters, denoted by $\{C_1, C_2, C_3\}$ are obtained as shown in Table 6.1. To measure the purity of clusters and validate experimental results, the F-measure has been used [46]. The F-measure is derived from *precision* and *recall,* the accuracy measures commonly employed in the field of information retrieval. The aforementioned three functions are shown by the following mathematical Eqs. 6.1 through 6.3 respectfully.

$$recall\left(N_p, C_q\right) = \frac{O_{pq}}{\left|N_p\right|} \tag{6.1}$$

$$precision\left(N_p, C_q\right) = \frac{O_{pq}}{\left|C_q\right|} \tag{6.2}$$

$$F\left(N_p, C_q\right) = \frac{2 * recall\left(N_p, C_q\right) * precision\left(N_p, C_q\right)}{recall\left(N_p, C_q\right) + precision\left(N_p, C_q\right)} \tag{6.3}$$

where O_{pq} is the number of members of actual (natural) class N_p in cluster C_q, N_p is the actual class of a data object O_{pq} and C_q is the assigned cluster of O_{pq}.

6.2.3 *Frequent Stylometric Pattern Mining*

Once clusters $\{C_1, \cdots, C_k\}$ are formed, the next step is to calculate the Writeprint of each cluster $C_i \in \{C_1, \cdots, C_k\}$. Pattern mining helps unveil the hidden association between different stylometric features. By feature items, we mean the discretized value of a feature, which is discussed in the following paragraph. Such frequently occurring patterns are captured by the concept of the frequent itemset [139] in a way similar to the one described in [64] and Chap. 5.

To extract frequent stylometric patterns from each cluster, the Apriori algorithm is applied [64]. The Apriori algorithm cannot be applied to numeric data. Therefore, it is important to split feature values into appropriate intervals. For this, each normalized frequency of a feature $F_a \in \{F_1, \cdots, F_8\}$ has been discretized into a set of intervals $\{\iota_1, \cdots, \iota_h\}$, called feature items. For example, frequencies from 0 to 1 are divided into 5 intervals i.e., $\{0.0{-}0.2, 0.21{-}0.4, 0.41{-}0.6, 0.61{-}0.8, 0.81{-}1.00\}$ and labeled as $\{X_1, X_2, X_3, X_4, X_5\}$ respectively. Table 6.2 shows the discretized form of the messages shown in Table 6.1.

A running example has been used to explain the proposed approach of writing style mining. Suppose there are three clusters, C_1 with messages $\{\mu_1, \mu_2, \mu_3, \mu_4\}$, C_2 with messages $\{\mu_5, \mu_6, \mu_7, \mu_8\}$, and C_3 containing messages $\{\mu_9, \mu_{10}, \mu_{11}, \mu_{12}\}$, as shown in Table 6.2. To calculate frequent stylometric patterns for each cluster, the user-defined *min_sup* = 0.5 has been assumed.

Table 6.3 Frequent stylometric patterns for clusters C_1, C_2, C_3

Cluster(C)	Frequent stylometric patterns (FP)
C_1	$\{X_2\}, \{Y_1\}, \{Z_1\}, \{X_2,Y_1\}, \{X_2, Z_1\}, \{Y_1, Z_1\},\{X_2, Y_1, Z_1\}$
C_2	$\{X_3\}, \{Y_1\}, \{Y_2\}, \{Z_2\}, \{X_3, Y_2\}, \{X_3, Z_2\}, \{Y_1, Z_2\}, \{Y_2, Z_2\}, \{X_3, Y_2, Z_2\}$
C_3	$\{X_2\}, \{Y_1\}, \{Z_3\} \{X_2, Y_1\}, \{Y_1, Z_3\}$

Table 6.4 Writeprints for clusters C_1, C_2, C_3

$WP(C_1)$	$\{X_2, Z_1\}, \{Y_1, Z_1\},\{X_2, Y_1, Z_1\}$
$WP(C_2)$	$\{X_3\}, \{X_3, Y_2\}, \{X_3,Z_1\}, \{Y_2,Z_1\}, \{X_3,Y_2,Z_1\}$
$WP(C_3)$	$\{Z_3\},\{Y_1,Z_3\}$

This means that a pattern P is frequent in C_i if at least 2 out of 4 e-mails (by truncating the decimal part) within a cluster C_i contain all feature items in P. The frequent stylometric patterns associated with each cluster are shown in Table 6.3. For instance, pattern $\{X_2, Y_1, Z_1\}$ is a frequent pattern in C_1 because at least 2 out of 4 e-mails of cluster C_1 contain this pattern. Similarly, $\{Y_1, Z_2\}$ is a frequent pattern in C_2 as it appears in 2 out of 4 e-mails of C_2. The lists of frequent stylometric patterns are shown in Table 6.3.

6.2.4 Writeprint Mining

A Writeprint is the disjoint set of frequent stylometric features. Therefore, the patterns that are shared by more than one cluster are dropped. For instance, in the given example, $\{X_2\}$ and $\{X_2, Y_1\}$ are shared by cluster C_1 and C_3, $\{Y_1\}$ is shared by all three clusters. $\{X_2, Y_1\}$ is shared by C_1 and C_3. Therefore, these patterns are deleted. The remaining frequent patterns constitute the unique Writeprints $WP(C_1)$, $WP(C_2)$, $WP(C_3)$, as shown in Table 6.4.

6.2.5 Identifying Author

In this section, the most plausible author for each cluster of anonymous messages C_j has been identified by comparing $WP(C_j)$ with the training samples $\{M_1, \cdots, M_n\}$. For each message in M_i, the stylometric feature items, denoted by $\{P(M_1), \cdots, P(M_n)\}$ have been extracted. If there are two or more samples, the average of the feature items across all messages in M_i is taken. The similarity between C_i and M_i is computed by using Eq. 6.4. The most plausible author is the suspect who has the highest score.

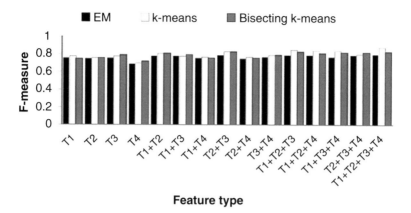

Feature type

Fig. 6.2 F-measure vs. Feature type (Authors = 5, Messages = 40) [149]

$$Score\left(M_i \approx WP\left(C_i\right)\right) = \frac{\sum_{j=1}^{\rho} support(MP_j | C_i)}{\left|WP\left(C_i\right)\right|} \qquad (6.4)$$

where $MP = \{MP_1, \cdots, MP_p\}$ is a set of matched patterns between $WP(C_i)$ and the message sample M_i of suspect S_i. A score is a real number within the range of [0, 1]. The higher the score means the greater the similarity between the cluster Writeprint $WP(C_i)$ and the message sample M_i. The author of the message sample M_i with the highest score for a cluster is the true author of that cluster. Suppose a message M_1 contains two patterns $\{X_3\}$ and $\{Y_1, Z_2\}$. Suppose the support of $\{X_3\}$ is 2 in cluster C_2 and the support of $\{Y_1, Z_2\}$ in cluster C_3 is 4. Using Eq. 6.4, as represented in Fig. 6.2, the score of cluster C_2 for M_1 is 0.4 and that of cluster C_3 is 4. Therefore, cluster C_3 is attributed to suspect S_1. The same process is repeated for the remaining two clusters as well. In the unlikely event that multiple suspects have the same highest score for a given cluster, the strategy discussed previously is applied.

6.3 Experiments and Discussion

The objective of these experiments is to evaluate the authorship identification accuracy of the proposed approach AuthorMinerSmall. Firstly, it has shown that clustering by stylometric features can be employed to group together the messages of an author. This is a two-step process. First, the randomly selected messages are clustered. Second, F-measure is used [152] to measure the similarity between the cluster solution and the true author labels. The higher the F-measure implies the better the cluster quality. F-measure has a range [0, 1].

Such cluster analysis experiments help answer the following questions. Which of the clustering algorithms performs best for a given message dataset? What is the relative strength of each of the four different types of writing style features? What is the effect of changing the number of authors on the experimental results? What is the effect of changing the number of messages per author on the cluster quality? To find the answer to the first question, three clustering algorithms have been applied, namely Expectation-Maximization (EM), k-means, and bisecting k-means. The cluster quality of the three algorithms is measured while all other parameters, e.g., stylometric features, number of authors, size of training data, etc., are kept constant. To answer the second question, the clustering over 15 different combinations of stylometric features is carried out. Next, the number of authors is changed while keeping the other parameters, e.g., feature set and size of training samples, constant. In the fourth set of experiments, the effects of changing the number of messages per author on the clustering result are checked.

A real-life dataset, Enron [148], real e-mails were used. The e-mail sets 200,399 contained e-mails from about 150 employees of Enron Corporation (after cleaning). H employees from the Enron e-mail dataset, representing h authors $\{A_1, \cdots, A_h\}$, are randomly selected. For each author A_i, x numbers of A_i's e-mails are selected; where h varies from three to ten while the value of x is selected from $\{10, 20, 40, 80,$ and $100\}$.

In the first set of experiments, 40 e-mails from each one of the five authors were selected. The results of the three clustering algorithms are shown in Fig. 6.2. The value of F-measure spans from 0.73 to 0.80 for EM, from 0.73 to 0.88 for k-means, and from 0.75 to 0.83 for bisecting k-means. The better results from k-means and bisecting k-means over EM (at least in this set of experiments) indicates that by knowing the number of clusters k, one can obtain better results. The result of k-means is better than bisecting k-means. Initially, these results were unexpected, but they were later validated after completing all sets of experiments. k-means performed better compared to bisecting k-means for up to 40 emails per author. By increasing the number of e-mails per author beyond 40, the accuracy of bisecting k-means starts increasing. This suggests that bisecting k-means is more scalable than EM and k-means.

The experimental results shown in Fig. 6.2 help measure the discriminating power of the different stylometric features. For this, 15 possible combinations of these features are used. Looking at the individual features, content-specific features (denoted by T_4) perform poorly while style markers (denoted by T_2) and structural features (denoted by T_3) produce the best clustering results. These two trends match previous stylometric studies [63, 65]. Overall, the best results are obtained by applying k-means on $T_1 + T_2 + T_3 + T_4$, i.e., the combination of all four types of features. By adding content-specific features to $T_1 + T_2 + T_3$, no noticeable improvement in the accuracy of EM and bisecting k-means has been seen. The main observations from the results are that the selected keywords are probably common among emails of the selected authors. Another important observation is that the performance of $T_2 + T_3$ is better than any other combination of two feature, e.g., $T_1 + T_2$ and $T_1 + T_3$.

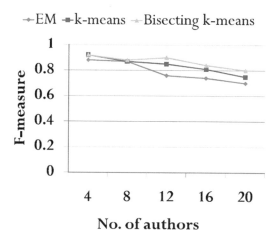

Fig. 6.3 F-measure vs. No. of authors (Messages = 40, Features = T1 + T2 + T3 + T4) [149]

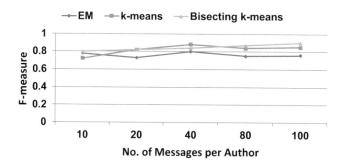

Fig. 6.4 F-measure vs. No. of messages per author (Authors = 5, Features = T1 + T2 + T3 + T4) [149]

In the second set of experiments, all four types of stylometric features, i.e., $T_1 + T_2 + T_3 + T_4$ were used and 40 messages per author are selected. As depicted in Fig. 6.3, the experiments were repeated for 4, 8, 12, 16, and 20 authors. The value of F-measure reached 0.91 for four authors using bisecting k-means. The accuracy of the three clustering algorithms dropped as more authors were added to the experiments. In the next set of experiments, the effects of the training size were evaluated by keeping the number of authors (five) and feature set $(T_1 + T_2 + T_3 + T_4)$ unchanged. As depicted in Fig. 6.4, the value of F-measure increased by increasing the number of messages per author. K-means and bisecting k-means achieved 90% purity for 40 messages per author while the results of EM were not consistent. Increasing the number of messages per author beyond 40 negatively affected the results of all the three algorithms. Among the three algorithms, the accuracy of EM dropped faster than the other two, and bisecting k-means is more robust compared to simple

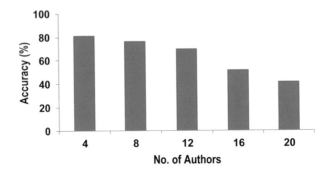

Fig. 6.5 AuthorMinerSmall: Accuracy vs. No. of authors

k-means. These results explain the relative behavior of these algorithms in terms of scalability.

The best level of accuracy was achieved by applying k-means over a combination of all four-feature types when e-mails per user were limited to 40. Bisecting k-means is a better choice when there are more authors and the training set is larger. Considering the topic of discussion, better results can be obtained by selecting domain-specific words carefully.

One way to improve results could be to identify author-specific keywords by applying content-based clustering to the e-mails of each author separately. The results of EM are insignificant and are hard to improve by parameter tuning. Next, the authorship identification accuracy of AuthorMinerSmall was evaluated. 40 text messages from each suspect were randomly selected, taking 36 out of the 40 messages from each suspect for training while the remaining 4 messages from each suspect were used for testing. Let n be the number of suspects. Subsequently, the 36 × n messages were clustered by stylometric features using k-means, and then each cluster of anonymous messages was matched with the remaining 4 × n messages with known authors. An identification is correct if AuthorMinerSmall can correctly identify the true author of an anonymous text message among the group of suspects.

Figure 6.5 depicts the authorship identification accuracy for AuthorMinerSmall with the number of suspects ranging from 4 to 20. When the number of suspects is 4, the accuracy is 81.18%. When the number of suspects increases to 20, the accuracy drops to 41.26%. Given that the training dataset is so small, an accuracy level above 70% is in fact very encouraging when the number of suspects is not too large. The computational complexity of the AuthorMinerSmall is based on two phases.

1. The computational complexity of the clustering phase depends on the clustering algorithm. For instance, it is $O(k \times |\Omega|)$ for k-means, where k is the number of clusters and $|\Omega|$ is the number of anonymous messages.

2. The computational complexity of Writeprint extraction phase is $O(|U| \times |C_1| \times k)$, where $|U|$ is the number of distinct stylometric feature items, I is the maximum number of stylometric features of any e-mail and $|C_i|$ is the number of anonymous messages in cluster C_i. I usually peaks at 2. For any test case of AuthorMinerSmall shown in Fig. 6.5, the total runtime is less than a minute.

The non-availability of enough training samples of potential suspects is one of the main limitations of the criminal investigation process. To address this issue, a method has been presented for the authorship identification of anonymous messages based on few training samples. The approach is primarily based on the intuition that clustering by stylometric features is a sensible method to divide text messages into different groups. We argue that the hypothesis is true based on these experiments on real-life e-mails. Moreover, it can be seen that using cluster analysis, an investigator can get a deeper insight into anonymous messages and learn about the potential perpetrators. In terms of feature patterns, the writing styles provide more concrete evidence than producing some statistical numbers.

The identification accuracy of AuthorMinerSmall for up to ten suspects is high while the accuracy above ten authors is low. This can be improved by tuning the parameters. For instance, selecting longer e-mails, increasing the number of stylometric features, and using sophisticated distance functions can help improve the accuracy score of the presented approach.

The current study suggests that content-specific keywords can be more effectively used for authorship identification in specific contexts, e.g., cybercrime investigations. More robust techniques need to be developed for the purposes of selecting more appropriate words from the given suspicious dataset. Another important research direction would be to identify an optimized set of stylometric features applicable in all domains. In many cases, the contents of the same message are written in more than one language. Therefore, addressing the issue of language multiplicity is important especially for future cybercrime investigations.

Chapter 7
Authorship Characterization

A problem of authorship characterization is to determine the sociolinguistic characteristics of the potential author of a given anonymous text message. Unlike the problems of authorship attribution, where the potential suspects and their training samples are accessible for investigation, no candidate list of suspects is available in authorship characterization. Instead, the investigator is given one or more anonymous documents and is asked to identify the sociolinguistic characteristics of the potential author of the documents in question. Sociolinguistic characteristics include ethnicity, age, gender, level of education, and religion [153].

In this chapter, the worst-case scenario of authorship characterization has been analyzed. In this worst-case scenario, the data from the sample population is not extensive enough to build a classifier. The proposed approach depicted in Fig. 7.1 first applies stylometry-based clustering to the given anonymous messages to identify major stylistic groups. The said approach suggests that clustering by stylometric features allows all messages written by one author to be clustered into one group. A group of messages is denoted by $C_i \in \{C_1, \cdots, C_n\}$. Following this, a model is trained on messages collected from the sample population. The developed model is employed to infer the characteristics of a potential author.

In this experiment, a blog dataset is used, as most bloggers voluntarily give away personal information on their blogs thus allowing characteristics to be determined. The selected bloggers need to be from the same class category that we want to infer in these experiments. For instance, to infer the gender of an author, we need to collect blog postings of male and female bloggers. Next, the Writeprint of each class category of the sample population is precisely modeled by employing the concept of frequent patterns [139]. The extracted Writeprints are then used to identify the class label of each message in a cluster C_i. The approach is used to predict two characteristics: gender, and region or location. In the remainder of this chapter, the term "online messages" has been used to refer to e-mail messages, blog postings, and chat logs.

F. Iqbal et al., *Machine Learning for Authorship Attribution and Cyber Forensics*, International Series on Computer Entertainment and Media Technology, https://doi.org/10.1007/978-3-030-61675-5_7

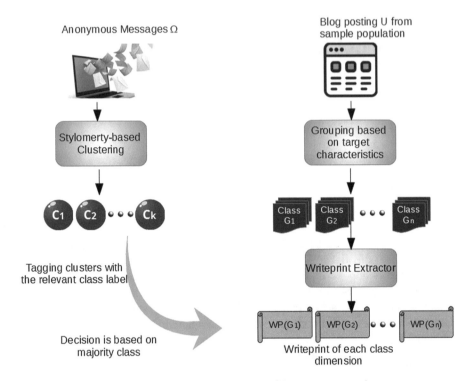

Fig. 7.1 AuthorCharacterizer: Inferring characteristics of the anonymous author

The contributions of this chapter are summarized as follows:

- **Characterization by frequent-pattern based Writeprint**: In traditional authorship studies, the characterization problem is addressed mostly by employing classifiers. This is the first work to use frequent-pattern based Writeprints to infer an author's characteristics. The combination of co-occurring stylistic features, known as the Writeprint, helps manifest the hidden association between the stylometric features.
- **Preliminary information:** Often, an investigator is provided only with a collection of anonymous suspicious messages and is asked to collect forensically relevant evidence from them. Clustering by stylometric features can be used to initiate the investigation process by identifying groups of stylistics; each group, intuitively, represents one suspect.
- **Small sample population:** The frequent-pattern based approach can be used even if the size of the sample population is small or if their sample messages are small. Sometimes, the data sets used are larger. A study by Koppel et al. [56] used the blogs of approximately 47,000 bloggers, across 1 year of posts, thus giving a data set that was much larger than the dataset used in most of these experiments.

- **A new category of characteristics:** Existing studies have investigated characteristics such as age, gender, educational level, and language background. A new dimension of authorship profiling called region or location has recently been introduced. Experiments on blog posts collected from bloggers in Australia, Canada, and the United Kingdom suggest that the proposed method can be employed to predict (with a certain level of accuracy) a suspect's region.

The problem of authorship characterization is defined as follows: Given a collection of anonymous online messages potentially written by some suspects, the task of an investigator is to identify the cultural and demographic characteristics of each suspect. It is assumed that there is no candidate list of potential suspects and no training data from the suspects. The investigator usually has access to some online messages with known authors who come from the population of the suspects. The sample messages can be collected from blog posts and social networks that explicitly disclose the authors' public profiles.

Definition 7.1 (Authorship characterization). Let Ω be a set of anonymous text messages potentially written by some suspects. The number of suspects may or may not be known. Both scenarios are addressed in this study. Let U be a set of online text documents, collected from the same population of suspects, with known authors' characteristics. The problem of authorship characterization is to first group the messages Ω into clusters $\{C_1, \cdots, C_k\}$ by stylometric features, then identify the characteristics of the author of each cluster C_j by matching the Writeprint extracted from the online text documents U.

7.1   Proposed Approach

To address the authorship problem stated above, a method called AuthorCharacterizer, has been proposed to characterize the properties of an unknown author of some anonymous messages. Figure 7.1 shows an overview of AuthorCharacterizer in three steps. Step 1 is to identify the major groups of stylometric features from a given set of anonymous messages n. Step 2 is to extract the Writeprints for different categories of online users from the given sample documents U. Step 3 is to characterize the unknown authors of n by comparing the Writeprints with n.

7.1.1   Clustering Anonymous Messages

Once all anonymous messages contained in *n* are converted into feature vectors using the vector space model representation technique, the next step is to apply clustering. For clustering, Expectation Maximization (EM), k-means, and bisecting kmeans clustering algorithms have been selected. These were chosen because k-means is more commonly used than other methods and because EM is the preferred choice if the number of clusters (the number of suspects in this case) is not known a priori. Bisecting k-means performs better than k-means in terms of accuracy. Clustering is applied on the basis of stylometric features, which results in a set of clusters $\{C_1, \cdots, C_k\}$. The only difference is that the number of clusters k is the number of categories identified for a characteristic. For instance, k = 2 (male/female) for gender, k = 3 (Australia/Canada/United Kingdom) for region or location.

7.1.2 Extracting Writeprints from Sample Messages

In this study, the blog posts collected and used were from blogger.com because this website allows bloggers to explicitly mention their personal information. Each blog post is converted into a set of stylometric feature items. Then these are grouped by the characteristics that are central in order to make inferences about the anonymous messages C_j. For example, if we want to infer the author gender of cluster C_j. the blog postings are divided into groups $G_1, \cdots, G_k$ by gender. Next, the Writeprints, denoted by $WP(G_x)$, are extracted from each message group G_x, by employing the method described earlier.

7.1.3 Identifying Author Characteristics

The last step infers the characteristic of the author of anonymous messages C_j by comparing the stylometric feature items of each message ω in C_j with the Writeprint $WP(G_x)$ of every group G_x. The similarity between ω and $WP(G_x)$ is computed using Eq. 7.1. Message ω is labeled with class x if $WP(G_x)$ has the highest Score ($\omega \approx WP(G_x)$). All anonymous messages C_j are characterized to label x that has the major class.

$$Score\left(\omega \approx WP\left(G_x\right)\right) = \frac{\sum_{x=1}^{\rho} support(MP_x | G_x)}{\left| WP\left(G_x\right) \right|} \qquad (7.1)$$

where MP = $\{MP_1, \cdots, MP_p\}$ is a set of matched patterns between $WP(G_x)$ and the anonymous message ω.

7.2 Experiments and Discussion

The main objective of these experiments is to evaluate the accuracy of the authorship characterization method, AuthorCharacterizer, based on the training data collected from blog postings. 290 stylometric features including the 285 general features and 10 genderspecific features were used. The 285 features described are gender-specific features, listed in Appendix II and described in [124].

The evaluation of AuthorCharacterizer is made up of three steps. In the first step, a small robot program is developed to collect blog posts with authors' profiles from a blogger website, grouping them by gender and location, and extracting the Writeprint of each group. To characterize the gender class, 50 posts/messages for each gender type were collected. Thus, if the total number of suspects is n, 50 × n × 2 blog posts in total are collected. The average size of each post is about 300 words. To characterize the location information, 787 posting from Australia, 800 posting from Canada, and 775 postings from the United Kingdom were collected. In the second step, the collected posts were clustered by stylometric features. 2/3 of the messages were used for training and 1/3 for testing. In the third step, the Writeprints were extracted from the training messages and were used to characterize the testing messages. Characterization of an anonymous message is correct if Author Characterizer can correctly identify the characteristic of the message.

Table 7.1 shows detailed experimental results for location identification. The actual accuracy is the percentage of records that are correctly characterized in a class. The weighted accuracy is normalized by the actual number of records whose class exceeds the total number of records. The sum is the sum of the weighted accuracy.

Table 7.1 Experimental result for location identification

No. of authors	Region/location	Accuracy (%)	W. accuracy (%)	Sum (%)
	AU	62.31	20.26	
4	CA	51.28	17.95	60.44
	UK	65.39	22.23	
	AU	46.99	15.04	
8	CA	62.00	21.7	50.18
	UK	39.52	13.44	
	AU	40.81	13.05	
12	CA	50.9	17.82	43.06
	UK	35.95	12.22	
	AU	39.98	12.79	
16	CA	49.16	17.21	43.21
	UK	38.6	13.21	
	AU	40.39	12.92	
20	CA	38.13	13.35	39.13
	UK	37.83	12.86	

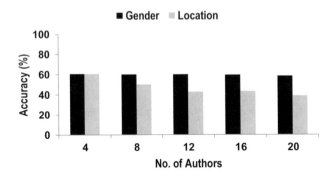

Fig. 7.2 Gender identification: Accuracy vs. No. of authors

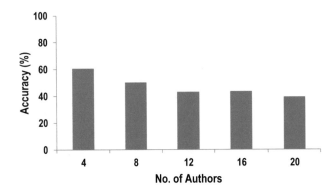

Fig. 7.3 Location identification: Accuracy vs. No. of authors

The accuracy scores of identifying the gender and location are depicted in Figs. 7.2 and 7.3, respectively. The accuracy stays almost flat at around 60% for gender and decreases from 60.44 to 39.13% as the number of authors increases for location. One apparent reason for this is the low number of classes in the case of gender characterization. The results suggest that the proposed frequent-pattern-based approach best fits the two-class classification problem. Another possible reason is the use of 11 gender-preferential features in addition to the general stylo-metric features in gender identification.

In this chapter, a technique has been presented for addressing the worst-case scenario of the characterization problem: when the size of the training data from the sample population is not sufficient. The proposed method has been evaluated using a blog dataset for two class dimensions: gender and location. In all experimental sets, this method identified the class labels correctly. Moreover, the notion of Writeprint, presented in the form of frequent patterns, is suitable for forensic purposes. Experimental results on real-life data suggest that the proposed approach, together with the concept of the frequent-pattern based Writeprint, is effective for identifying the author of online messages and for characterizing an unknown author.

Chapter 8
Authorship Verification

In the previous chapters, methods to address two authorship problems, i.e., authorship identification and authorship characterization, were proposed. This chapter discusses the third authorship problem, called authorship verification. The proposed approach is applicable to different types of online messages, but in the current study, the focus is on e-mail messages.

A problem of authorship verification is to ascertain whether a given suspect is the true author of a disputed textual document. Some researchers define authorship verification as a similarity detection problem, especially in cases of plagiarism. In such a situation, an investigator needs to decide whether the two given objects are produced by the same entity. The object in question can be a piece of code, a textual document, or an online message. More importantly, the conclusion drawn needs not only to be precise but to be supported by strong presentable evidence as well.

The problems of authorship attribution and characterization, discussed in previous chapters, are relatively well defined, but authorship verification is not. Sometimes, it is considered as a one-class text classification problem while elsewhere it can be considered a two-class classification problem. Some studies address the problem by determining the dissimilarities between the writing styles of the suspect and a pseudo-suspect. In addition to this, the metrics employed for measuring verification results vary from study to study. The measures include ROC curves [128], precision, recall [62], p-test, and t-test [63].

In this chapter, the problem of authorship verification is formally defined, and an authorship verification framework is proposed for e-mails. This method is primarily based on the speaker recognition evaluation (SRE) framework developed by the National Institute of Standards and Technology (NIST) [154], which has proven to be very successful in the speech processing community. The SRE framework evaluates the performance of detection systems in terms of minDCF, false positive, and false negative alarms represented by employing a detection error trade-off (DET) curve, a deviant of the receiver operating characteristic (ROC) curve (see details in Sect. 8.2 below).

© The Editor(s) (if applicable) and The Author(s), under exclusive license to
Springer Nature Switzerland AG 2020
F. Iqbal et al., *Machine Learning for Authorship Attribution and Cyber
Forensics*, International Series on Computer Entertainment and Media
Technology, https://doi.org/10.1007/978-3-030-61675-5_8

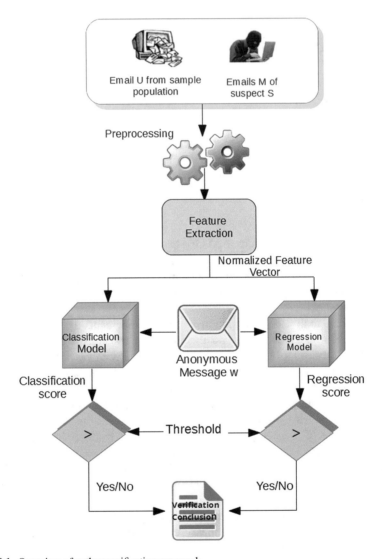

Fig. 8.1 Overview of author verification approach

The overview of the proposed approach is shown in Fig. 8.1 for two e-mail datasets. One is collected from a very large sample population denoted by U, and the other is confiscated from a potential suspect S. After the necessary preprocessing steps (cleaning, tokenization, stemming, and normalization), each e-mail is converted into a vector of stylistics or stylometric features (discussed in Chap. 5). Classification and regression techniques have been applied to both the datasets. In each thread of techniques, the datasets are further divided into two subsets, the training, and the testing sets. Two different models, one each for suspect S, called the hypothesized author and the alternate hypothesis, are trained and validated.

The given anonymous e-mail is evaluated using the two models in both regression and classification threads. Unlike the usual classification where the decision is made solely based on matching probability, here the decision to verify the author is based on the threshold defined for the hypothesis testing. The threshold is calculated by varying the relative number of false positives and false negatives, depending upon the nature of the perceived application of the system. The accuracy of the system is judged in terms of EER, represented by the DET curve, and the minDCF, as using only EER can be misleading [56].

Our experimental results on real-life data show that the proposed verification method has the following main contributions.

- *Adopting NIST Speaker Recognition Framework:* This is the first study that successfully adopted the NIST's SRE framework for addressing the issue of authorship verification of textual content with an e-mail dataset. Different classification and regression methods have been used to achieve an equal error rate of 17% and minDCF equal to 0.0671 with the SVM-RBF (Support Vector Machine—Radial Basis Function).
- *Employing regression for binary classification:* Regression functions, normally used for predicting numeric attributes (class labels), are employed for making a binary decision about whether or not a suspect is the author of a disputed anonymous document. It is evident from the experimental results that SVM with RBF kernel produced the best verification accuracy with the lowest minDCF value as compared to the classifiers used.
- *Proposing new error detection measures for authorship verification:* To measure the performance of most detection tasks, a ROC curve is traditionally used, where false alarms are plotted against the correct detection rate. Using this approach, it is hard to determine the relative ratio of both types of errors, which is crucial in a criminal investigation. The DET curve employed in this study can better analyze the exact contribution of both the false positive and false negative values. The use of EER is augmented using minDCF to gauge the framework's accuracy.

8.1   Problem Statement

Given a set of sample e-mails of a potential suspect S and an e-mail dataset U collected from a very large population of authors, the task of an investigator is to verify whether or not the disputed anonymous e-mail c_o is written by the suspect S. Mathematically, the task of author verification can be termed as a basic hypothesis test between

H_0: ω is written by the hypothesized author S
    and
H_1: ω is not written by the hypothesized author S.

The optimum test to decide between these two hypotheses is a likelihood ratio test given by

$$\frac{p\left(\mu|H_0\right)}{p\left(\mu|H_1\right)} \geq \varnothing$$

accepting H_0, otherwise reject H_0 (accept H_1) where $p(\mu|H_i)$, $i = 0, 1$ is the probability density function for the hypothesis H_i evaluated for the observed e-mail c_0 and $\varnothing$ is the decision threshold for accepting or rejecting H_0. The basic goal is to find techniques for calculating the two likelihood functions $p(\mu|H_0)$ and $p(\mu|H_1)$. The author-specific model H_0 is well defined and is built using e-mails written by the hypothesized author, while the model H_i is not well defined, as it must represent the entire space of the possible alternatives to the hypothesized author.

In order to define the H_1 model, the techniques used in the speaker verification literature are borrowed. Two main approaches are used for alternative hypothesis modeling in speaker recognition research. The first approach is to use a set of other author models to cover the space of the alternative hypothesis. This set of authors is called the cohort or background authors. Given a set of N background author models $\lambda_1, \lambda_2, \cdots, \lambda_N$, the alternative hypothesis model is represented by

$$p\left(\mu|H_1\right) = f\left(p\left(\mu|\lambda_1\right), p\left(\mu|\lambda_2\right), \cdots, p\left(\mu|\lambda_N\right)\right)$$

where f (.) is some function, such as the average or maximum, of the likelihood values from the background author set. The selection, size, and combination of the background authors can be the subject of further research.

Another approach is the alternative hypothesis modeling approach in which a model is developed on sample documents collected from a very large number of individuals. This is called the universal background model (UBM) in the speech processing community. The same approach has been adopted for online textual documents. Given a collection of e-mail samples from a very large number of authors, a single model is trained to represent the alternative hypothesis. The main advantage of this approach is that a single author-independent model can be trained once for a particular task and then used for all hypothesized authors in that task. Two types of errors can occur in the author verification system, namely false rejection (rejecting a valid author) and false acceptance (accepting an invalid author). The probabilities of these errors are called false rejection probability P_{fr} and false alarm probability P_{fa}, where both types of error depend on the value of user-defined threshold 0. It is, therefore, possible to represent the performance of the system by plotting P_{fa} versus P_{fr}, the curve generally known as DET curve in the speech processing community.

In order to judge the performance of the author verification systems, different performance measures can be used. The two main measures, namely Equal Error Rate (EER) and Detection Cost Function (DCF) are borrowed from the speech processing community. The EER corresponds to the point on the DET curve where

$P_{fa} = P_{fr}$. Since using only EER can be misleading [56], the DCF is used in conjunction with EER to judge the performance of the author verification system. The DCF is the weighted sum of miss and false alarm probabilities [154]. The DET curve is used to represent the number of false positives versus false negatives. The point on the DET curve where the number of both false alarms become equal is called EER. The closer the DET curve to the origin, the lower the EER is and thus the better the system is.

$$DCF = C_{fr} \times P_{fr} \times P_{target} + C_{fa} \times P_{fa} \times \left(1 - P_{target}\right)$$

The parameters of the cost function are the relative costs of detection errors, C_{fr} and C_{fa} and the a priori probability of the specified target author, P_{target}. In this method, the parameter values, as specified in the NIST's SRE framework, are used. These values are

$$C_{fr} = 10, C_{fa} = 1 \text{ and } P_{target} = 0.01.$$

The minimum cost detection function (minDCF) is redefined as the minimum value of $0.1 \times$ false rejection rate $+ 0.99 \times$ false acceptance rate'. Since it is primarily dependent on the false acceptance rate and false rejection rate and has nothing to do specifically with speech, it can be used for authorship verification as well. It is in conformance with the forensic analysis and strictly penalizes the false acceptance rate as it would implicate an innocent person as the perpetrator.

8.2 Proposed Approach

This chapter addresses authorship verification as a two-class classification problem by building two models; one from the e-mails of the potential suspect, and the other from a very large e-mail dataset belonging to different individuals called the universal background model. To train and validate the two representative models, techniques from the SRE framework [154] are borrowed. The framework was initiated by the National Institute of Standards and Technology. The purpose of the SRE framework is not only to develop state-of-the-art frameworks for addressing the issues of speaker identification and verification but also to standardize and specify a common evaluation platform for judging the performance of these systems as well.

The evaluation measures such as DCF, minDCF, and EER that are used in the SRE framework are more tailored to forensic analysis compared to simple ROC and classification accuracies. Another reason for borrowing ideas from the speaker recognition community is that this area has a long and rich scientific basis with more than 30 years of research, development, and evaluation [154]. The objective of both authorship and speaker verification is the same, i.e., to find whether or not a particular unknown object is produced by a particular subject. The object in this case is the anonymous e-mail whereas, in the case of speaker verification, it is the speech segment. The subject is the speaker in cases of speaker verification whereas it is the author in the case of authorship verification.

As depicted in Fig. 8.1, the proposed method is a two-step process: model development and model application. In the first step, the given sample data is used to develop and validate the classification model. Next, the disputed anonymous message is matched with the model to verify its true author. Prior to model development, the given sample messages are converted into feature vectors. To confirm whether a given anonymous e-mail co belongs to the hypothesized author (or suspect S) or not is based on the scores produced by e-mail co during the classification process and threshold 0. The threshold is defined by the user and is employed for taking a binary decision. As described in the following paragraphs, two approaches are used for the binary classification of e-mails.

8.2.1 Verification by Classification

In this approach, the e-mails in the training set corresponding to the hypothesized author, and belonging to the sample population, are nominally labeled. During the testing phase, a score is assigned to each e-mail based on the probability assigned to the e-mail by the classifier. The scores calculated for the true author, the 'imposters' are evaluated for false acceptance and false rejection rates through a DET plot. Three different classification techniques, namely Adaboost [155], Discriminative Multinomial Naive Bayes (DMNB), [156], and Bayesian Network [157] classifiers, are used. Most of the commonly used classification techniques including the one employed in the current study are implemented in the WEKA toolkit [94]. WEKA is described in Chap. 3.

8.2.2 Verification by Regression

Authorship verification is conceptually a classification problem but, in this case, there is a need to make a binary decision to determine whether or not the message being tested belongs to the potential suspect. Since the decision is made based on the similarity score assigned to the e-mail being tested, regression functions are employed to calculate the score. Three different regression techniques including linear regression [94], SVM with Sequential Minimum Optimization (SMO) [158], and SVM with RBF kernel [159] are used. Regression scores have been used for the true authors and for the impostors to calculate equal error rate and minimum detection cost function.

An integer is assigned to the e-mails of the true author and those belonging to the 'imposters'. For instance, +10 is assigned to the hypothesized author's e-mails and −10 to e-mails of the target population. When these integers have been assigned, the regression function assigns a value generally between +10 and −10 to the disputed anonymous e-mail. The decision of whether or not the anonymous e-mail belongs to the hypothesized author is based on the resultant score and the user-defined threshold 0.

Setting the threshold too low will increase false alarm probability whereas setting it too high will have a high miss probability (false rejection rate). In order to decide the optimal value of the threshold and to judge the performance of the verification system, the variation of the false alarm rate with the false rejection rate is plotted. The curve is generally known as the detection error trade-off curve, which is drawn on a deviate scale [154]. The closer the curve to the origin, the better the verification system is. The point on the curve where the false alarm rate equals the false rejection rate is called the equal error rate.

8.3 Experiments and Discussion

To evaluate the implementation, experiments are performed on the Enron e-mail corpus made available by MIT. First, a universal background model is created from the entire Enron e-mail corpus. This is an author-independent model and it is used as the basis for making the decision of whether the e-mail in question belongs to the suspected author. A separate model is created for each author. For this, 200 e-mails per author are used. The decision of whether an e-mail being tested belongs to the hypothesized author or not is based on the difference of similarity of the e-mail to the author-independent model and to the hypothesized author model. Based on this similarity metric, a score is assigned to the disputed e-mail. For the evaluation of the classification methods, the widely used tenfold cross-validation approach is employed, reserving 90% for training, and 10% for testing. The reason for this is to avoid any biases during the evaluation process and to judge the classification method over the entire database.

One of the performance measures used in the SRE framework is to calculate the equal error rate [154]. The EER is calculated by taking two types of scores as input, namely the true author score and the false author score, which in turn are calculated by the classification methods applied over the test dataset.

8.3.1 Verification by Classification

Figure 8.2 depicts the DET plot of the classification results of one author, randomly selected from the database. Usually, the closer the DET curve to the origin, the lower the EER is and thus the better the system is. The point on the DET plot which gives the minimum cost detection function is marked with a small circle on each curve.

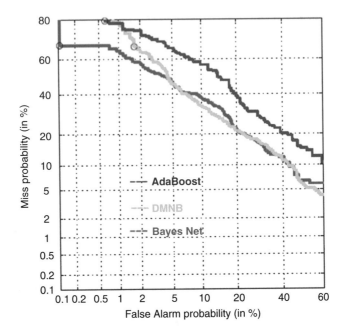

Fig. 8.2 DET for author verification using classification techniques

The DET curve plotted for Bayesian Network (BayesNet) is more consistent and indicates better results both in terms of EER and minimum cost detection function with less complexity. The values of minDCF for both DMNB and AdaBoost are comparable, however, the performance of DMNB in terms of EER is closed to BayesNet. The performance gap between the two classifiers is consistent in most of the experiments.

8.3.2   Verification by Regression

Figure 8.3 shows the typical DET plot of one of the randomly selected authors from the database. It was constructed using the scores obtained from the three regression techniques as described above. The DET curve indicates that the regression approach usually produces better results in terms of EER and minDCF compared to the classification approach. The regression approach via SVM with RBF kernel with EER 17.1% outperformed linear regression (with EER = 19.3%) and SVM-SMO (with EER = 22.3%). The same performance tendency can be seen in minDCF values as well (see the last row of Table 8.1).

DET curves for linear regression and SVM-SMO are running neck to neck starting with the highest value of false negative.

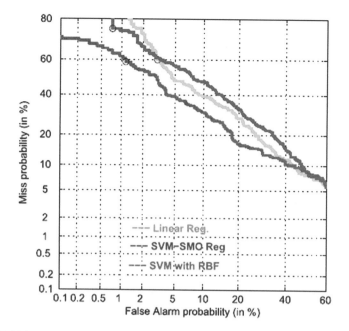

Fig. 8.3 DET for author verification using regression techniques

Table 8.1 Verification scores of classification and regression methods

	Classification			Regression		
Verification	A.Boost	DMNB	Bayes	SVM-SMO	Lin. Reg	SVM-RBF
EER(%)	22.4	20.1	19.4	22.3	19.3	17.1
minDCF	0.0836	0.0858	0.0693	0.0921	0.0840	0.0671

SVM with RBF kernel produced the best verification accuracy with the lowest rninDCF value. These results suggest that regression techniques are more suitable for addressing verification problems than classifiers, which perform better in attribution issues. However, the same assumption may not always be true, and the result may change depending on the dataset as well as the feature set used.

This chapter has presented the problem of e-mail authorship verification and has proposed a solution by adopting the NIST speaker verification framework and the accuracy measuring methods. The problem has been addressed as a two-class classification problem by building two models, one from the e-mails of the potential suspect and the other from a very large e-mail dataset belonging to different individuals, called the universal background model. Experiments on a real-life dataset produced an equal error rate of 17% by employing support vector machines with RBF kernel, a regression function. The results are comparable with other state-of-the-art verification methods. Building a true 'universal' background model is not an easy task, due to the non-availability of sufficient sample e-mails. The style variation of the same suspect with the changing state of mind and the context in which he writes may affect his representative model.

Chapter 9
Authorship Attribution Using Customized Associative Classification

In this chapter, Associative Classification (AC) [139] is employed, based on association rule discovery techniques, for authorship identification. The developed classification model consists of patterns that represent the respective author's most prominent combinations of writing style features.

There are many different implementations of AC, namely Classification based on Associations (CBA) [160], Classification based on Predictive Association Rules (CPAR) [161], Classification-based on Multiple Association Rules (CMAR) [162], and Multi-class Classification based on Association Rule (MCAR) [163]. Given the need to accurately quantify the match between the various author's writing styles and the anonymous e-mail, the main focus of this chapter is on CMAR [162]. This variation on AC uses a subset of rules as opposed to a single best rule, to determine which class, or author in this case, is the best match. Below are some of the pertinent contributions of this chapter.

- To our knowledge, this is the first application of AC to the authorship attribution problem; the experimental results on real-life data endorse the suitability of the presented approach.
- Association rule mining in associative classification is different than traditional association rule mining; the former investigates the associativity of features with one another as well as with the target predetermined classes, whereas the latter is limited to the analysis of the interdependence between features and does not associate them all to a target class. Therefore, extracted association rules reveal feature combinations that are relevant in distinguishing one author from another in authorship identification.
- Each instance in a classification model shows the features that are related, not only to each other but to the class label as well. As a result, the proposed method builds a concise and representative classifier that can serve as admissible evidence to support the identification of the true author of a disputed discourse.

F. Iqbal et al., *Machine Learning for Authorship Attribution and Cyber Forensics*, International Series on Computer Entertainment and Media Technology, https://doi.org/10.1007/978-3-030-61675-5_9

9.1   Problem Statement

Let S be the group of suspected authors of an anonymous e-mail e. Let E_i be a relatively large collection of e-mails written by suspect $S_i \in S$. The problem of authorship attribution is to find the most plausible author S_a from the suspects S, whose e-mail collection E_a best matches with the stylometric feature items in the malicious e-mail e. Intuitively, an e-mail collection E_i matches e if E_i and e share similar patterns of writing style features in strongly representative combinations. The primary objective of cyber forensics investigators is to automatically and efficiently model the patterns, or Writeprint, of each suspect. They can then present such patterns as convincing evidence identifying the author of the malicious e-mail e.

In terms of associative classification, it is necessary to identify what exactly comprises a Writeprint. Specifically, the objective is to extract rules derived from patterns that strongly and uniquely represent the writing style of each suspect S_i, but which do not embody the writing style of any other suspect S_j, where $i \neq j$. The rest of this section will discuss the pre-processing of e-mails and formally define the notions of frequent patterns, classification rules, and similarity metrics between an e-mail collection E_i and an anonymous e-mail e.

9.1.1   Extracting Stylometric Features

For each e-mail, the headers and appended forward or reply content are removed. E-mails with less than a few sentences or unrelated text attached are not included, as they would not contain sufficient information about the author's writing style.

Numerical features are normalized to a value between 0 and 1, and then each normalized feature is discretized into a set of intervals, for example, [0–0.25], [0.25–0.5], [0.5–0.75], [0.75–1], based on equal-frequency discretization, where each interval contains approximately the same number of records. Each interval is designated as a feature item. The subsequently normalized frequency of features is then matched with these intervals. Intuitively, the writing style of a collection of e-mails E_i written by suspect S_i is a combination of stylometric feature items that frequently occur in e-mails E_i. These frequently occurring patterns are modeled with the concept of the frequent pattern [139] or frequent stylometric pattern described as follows in [64].

Definition 9.1 (Frequent stylometric pattern). Let E be an e-mail collection. Let sup(F) be the e-mails in E that contain the stylometric pattern $F \subseteq V$. A stylometric pattern F is a frequent stylometric pattern in E if sup(F) $\geq$ min_sup, where the minimum support threshold min_sup is a positive real number provided by the user.

The writing style of a suspect S_i is therefore represented as a set of frequent stylometric patterns, denoted by $FP(E_i) = \{F_1, \cdots, F_k\}$, extracted from the set of e-mails E_i. These patterns are used to derive a high-quality class association rule list that consists of frequent stylometric patterns by means of pruning and ranking.

Example 9.1 For two stylometric patterns $\{X_1, Y_2\}$, $\{X_1, Y_1, Z_2\}$ in E, $\text{sup}(\{X_1, Y_2\}) = 5$ whereas $\text{sup}(\{X_1, Y_1, Z_2\}) = 3$. If min_sup = 4, $\{X_1, Y_2\}$ is a frequent stylometric pattern in E whereas, $\{X_1, Y_1, Z_2\}$ is not a frequent stylometric pattern in E.

9.1.2   Associative Classification Writeprint

The task of classification has been described in Chap. 5. A more recent approach is to explore strong relationships between specific sets of objects features and their class labels; frequent patterns in records with the same class label can then be used to infer the class of other records with similar patterns. The important advantage of using AC over classical classification approaches is that the output of an AC algorithm is represented by simple If-Then rules which are easy and intuitive to interpret and understand.

Let S be a finite set of distinct class labels, each representing a suspect in this context. A training data set is a set of e-mails, each with an associated class label S_i $\in$ S. A classifier is a function that maps an e-mail to a class $S_i \in$ S. AC is the process of discovering Class Association Rules (CAR) that capture the relationship between the combinations of stylometric features and the suspects. Specifically, the antecedent of a CAR contains a combination of stylometric features and the consequent of a CAR is a suspect. The support and confidence of a CAR have to pass the minimum support and minimum confidence thresholds specified by the operator. The notion of CAR is formally defined as follows.

Definition 9.2 (Support of a rule). Let $A \rightarrow B$ be an association rule, where $A \subseteq V$, $B \in S$. The support of $A \rightarrow B$, denoted by $\text{sup}(A \rightarrow B)$, is the percentage of e-mails in E containing $A \cup B$.

Definition 9.3 (Confidence of a rule). Let $A \rightarrow B$ be an association rule, where $A \subseteq V, B \in S$. The confidence of $A \rightarrow B$, denoted by $\text{conf}(A \rightarrow B)$, is the percentage of e-mails containing B that also contains A.

Definition 9.4 (Class association rule (CAR)). A class association rule has the form $A \rightarrow B$, where $A \subseteq V, B \in S$, $\text{sup}(A \rightarrow B) \geq$ min_sup, and $\text{conf}(A \rightarrow B) \geq$ min_conf, where min_sup and min_conf are the minimum support and minimum confidence thresholds specified by the user.

For example, if 90% of suspect S_i's e-mails contain 3 paragraphs, then the confidence of rule conf(3 paragraphs→S_i) is 90%. This rule can be used to classify future records that match this pattern.

The minimum support threshold is used to avoid noise. Typically, AC finds the complete set of CARs that pass the user-supplied minimum support and confidence thresholds. When a new record requires classification, the classifier will select the matching rule with the highest confidence and support and use it to predict the class label. Recently proposed AC techniques will prune and rank rules and sometimes even use multiple rules to predict the class label of an unknown record as there are situations in which the single best rule may not be the most intuitive or even most appropriate choice. Many studies show that AC is intuitive, efficient, and effective.

Authorship attribution requires special attention when it comes to using AC techniques to obtain the best results; with multiple distinct classes and the need to consider much more than simply the strongest class, it becomes evident that a classifier should consider as much information as possible. Example 9.1. demonstrates why a single matching rule may not always be the best choice.

Example 9.1 Suppose we want to find the author of an anonymous e-mail with feature items $(2, 5, 8)$. The top three most confident rules matching the e-mail are as follows:

> Rule R1:2 → Suspect 0 (support: 33%, confidence: 90%)
> Rule R2:5 → Suspect 1 (support: 67%, confidence: 89%)
> Rule R3:8 → Suspect 1 (support: 50%, confidence: 88%)

Most AC techniques that select the most confident rule would classify this e-mail as belonging to Suspect 0, but a closer look suggests that this decision has been made with no regard to the rest of the rule list. All three rules have similar confidence levels but both R2 and R3 have higher support which means that the values of those features were found more often in the training data set for Suspect 1. Suspect 1 is, therefore, a more intuitive choice and the proposed algorithm should take this into account. Situations like this make it clear that in order to make a reliable and accurate prediction, especially when the result could mean the difference between guilty and innocent, an aggregate measure analysis based on multiple rules leads to a higher quality classification.

Many studies have presented ways of greatly diminishing the number of class association rules so as to improve efficiency, given that usually only the strongest rule would be used for classification anyway. This approach uses multiple rules [162] and so it is important not to prune or discard too much information. In general, rules with low support and confidence are pruned or outranked by more powerful rules, regardless of their class association. This means that a given author may be assigned to an unknown e-mail simply because he/she has a stronger Writeprint than the true author and not based on a normalized measure of similarity. This would be the equivalent of identifying a matching fingerprint against two samples: one with a full print and another just with a partial print. The full print has more potential to match or to mismatch the unknown print, whereas the partial sample, even if it matches the unknown print very well, could still be discarded as its potential to fully match the unknown print is inherently lower. Once a set of CAR is discovered, the next step is to remove common rules among the suspects because the objective is to identify the combinations of stylometric feature items that can uniquely identify the author from a group of suspects.

After pruning the common rules, the remaining list of CARs, denoted by WPCAR, encapsulates the Writeprints of the suspects.

Definition 9.5 (CAR Writeprint). The Writeprint of a suspect S_i, denoted by WP (S_i), is the set of rules in WPCAR with the form $A \rightarrow S_i$.

Our proposed notion of CAR Writeprint is different from the traditional authorship Writeprint in previous works [110]. The first distinction is that the feature item combination that composes the Writeprint of a suspect S_i is generated dynamically based on the unavoidable patterns present in their e-mails E_i. This flexibility allows us to concisely model the Writeprint of different suspects by using various feature item combinations. Secondly, every rule in this notion of Writeprint captures a writing pattern that can only be found in a single suspect's collection of e-mails. A cyber forensic investigator could then precisely point out a matched pattern in the malicious e-mail to support his/her conclusion of authorship identification. In contrast, a traditional classifier, such as a decision tree, might use the same feature set to capture the Writeprint of different suspects. It would be dangerous for the classifier to capture common writing patterns and use them as evidence that points to multiple authors; drawing a legal conclusion based on ambiguous evidence is problematic for obvious reasons. The proposed notion of Writeprint avoids this ambiguity and, therefore, produces more reliable and convincing evidence.

The removal of common patterns certainly improves the quality of the derived Writeprint, especially for the purpose of evidence collection. However, one must understand the advantages as well as the disadvantages inherent in this technique. If there are a large number of suspects, it is entirely possible for one author's Writeprint to completely intersect with the union of the other authors' Writeprints, leaving them without any Writeprint at all. This could happen if the set of common rules is equivalent to the total set of rules for one class.

9.1.3   Refined Problem Statement

The problem of authorship attribution by multiple class association rules can be refined into three sub-problems:

1. To discover the CAR Writeprint WP (S_i) of each suspect S_i from the training e-mail sets E,
2. To identify the author of the malicious e-mail e by matching e with WP(E_1), ···, WP(E_n), and
3. To extract convincing evidence for supporting the conclusion on authorship. Any evidence must be of high enough quality to convince a judge and jury in a court of law.

These three sub-problems outline the challenges in a typical investigation procedure and reflect the use of associative classification in this process.

To solve sub-problems (1) and (2), rules are mined by extracting the frequent patterns and list of class association rules from the training set E while ranking and pruning them to build a representative final CAR list. For subproblem (3), the matching group of rules with the best score serves as evidence for supporting the conclusion.

9.2   Classification by Multiple Association Rule for Authorship Analysis

In this section, a novel machine learning strategy is presented that utilizes the concept of frequent stylometric patterns and AC to address the three authorship attribution problems. Section 9.2.1 first presents a method to extract frequent stylometric patterns and class association rules. Section 9.2.2 presents the procedure for pruning the irrelevant or common rules that are shared among multiple suspects. Finally, Sect. 9.2.3 discusses how to use the rules to determine the most plausible author of an anonymous e-mail.

9.2.1   Mining Class Association Rules

A CAR list is compiled by mining a training data set to find the complete set of rules passing user-supplied minimum support and confidence thresholds. This is comparable to any frequent pattern mining or association rule mining task. Classification by CMAR [162, 164] forms the basis of the AC methods described in this study. The algorithm, which is used to mine rules is a variant of the FP-growth algorithm [165]. Making use of efficient tree structures [164], first a partial support tree and then a total support tree, database scans are minimized and there is no need to generate candidate sets. The benefits of this method are especially apparent when processing large datasets with a low support threshold and a large number of features. This situation is commonly seen in authorship attribution problems. Furthermore, accuracy is generally better when low support and confidence thresholds are used, making this choice of algorithm suitable.

The concept of class association rules mining has been illustrated below with examples. Refer to [162] for more details about the algorithm. Consider the training set in Table 9.1. Setting the support threshold to 2 and confidence to 50%, the algorithm extracts class association rules as follows.

1. The training set T is scanned, retrieving the set of feature items that pass the minimum support threshold. The set $F = \{a_1, b_1, c_2, d_3\}$ is called a frequent

Table 9.1 Training set

E-mail	Feat.A	Feat.B	Feat.C	Feat.D	Auth.ID
1	a2	b1	c2	d1	A
2	a1	b1	c2	d3	B
3	a3	b2	c1	d2	A
4	a1	b1	c3	d3	C
5	a1	b1	c2	d3	C

itemset, as each element in the set appears at least twice. All other feature items appear only once and are pruned.

2. The feature items in F are then sorted in descending order to become F = $\{b_1, a_1, c_2, d_3\}$. The database is then scanned again to construct an FP-tree as shown in Fig. 9.1a. A FP-tree is a prefix tree with regard to the F-list. For each tuple in the training dataset, feature items appearing in the F-list are extracted then sorted accordingly. For example, for the first tuple, $\{b_1, a_1\}$ is extracted and inserted in the tree as the left-most branch. The author ID is appended to the last path node. For efficiency, training data tuple sets always share prefixes. For example, the second tuple carries feature items $\{b_1, c_2, a_1\}$ in the F-list and shares a common prefix b_1 with the first tuple. The sub-path with the left-most branch is therefore shared. All nodes sharing the same feature items are queued starting from the described header table.

3. The set of class association rules based on the F-List can be branched into four subsets with no overlap:

   – Those with d_3;
   – Those with a_1 but not d_3;
   – Those with c_2 but not d_3 or a_1; and
   – Those with b_1 exclusively.

   These subsets are discovered iteratively one at a time.

4. Finding the subset of rules that have d_3, the algorithm involves traversing nodes that have feature item d_3 and looking upwards to construct a d_3-projected database, which contains three tuples: $\{b_1, c_2, a_1, d_3\}$: C, $\{b_1, c_2, d_3\}$: C and d_3: A. Given that all tuples containing d_3 are included, the challenge of finding all frequent patterns with d_3 in the entire training data set can be simplified to mining patterns in a d_3-projected database. Passing the support threshold, b_1 and c_2 are frequent feature items in the d_3-projected database. d_3 does not count as a local frequent attribute because, in a d_3-projected database, d_3 is present in every tuple and is therefore trivially frequent. The projected database can be mined recursively by constructing FP-trees and other projected databases, as described in detail by Han et al. [165]. In d_3-projected databases, b_1 and c_2 always appear together, they are both sub-patterns of $b_1 c_2$ and have the same support count as $b_1 c_2$. The rule R: $b_1 c_2 d_3 \rightarrow$ C with support 2 and confidence 100% is generated

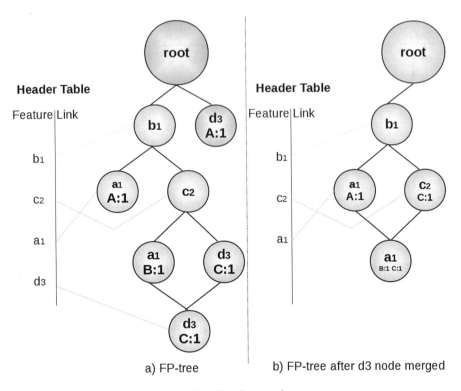

a) FP-tree   b) FP-tree after d3 node merged

Fig. 9.1 (**a**) FP-tree [166]. (**b**) FP-tree after d3 node merged

based on author distribution. After processing all rules that include d_3, those nodes can be merged into their parent nodes. This means that any class association registered in any d_3 node is registered with its parent node, effectively shrinking the FP-Tree to what is shown in Fig. 9.1b. This operation is performed while the d_3-projected database is being built.

This process is then repeated for the remaining subsets of rules.

9.2.2 Pruning Class Association Rules

Class association rule mining can generate an enormous number of rules; it is advantageous and rather simple to prune redundant or noisy rules in order to build a concise yet high-quality classifier. The associative classification variant [162] used in this study is modified from the rule ordering protocol called CBA [174]. The final rule list is ordered according to three ranking conditions. Given rules R1 and R2, R1 will be assigned a higher rank than R2, denoted by R1 ≻ R2, if and only if

1. Conf(R1) > conf(R2),
2. Conf(R1) = conf(R2) but sub(R1) > sup(R2), or
3. Conf(R1) = conf(R2) and sup(R1) = sup(R2) but |ant(R1)| < |ant(R2)|.

Rule R1: P → c is described as a general rule of rule R2: P' → c', if and only if P ⊆ P'.

The first round of pruning uses ambiguous and high confidence rules to prune more specific and lower confidence rules. Given two rules R1 and R2, where R1 is a general rule with regard to R2. CMAR [162] will prune R2 if R1 also has a higher rank than R2. The theory is that general rules with high confidence are more representative than more specific rules with low confidence, so we can prune the specific and low confidence ones. However, we will see that this is not an ideal behavior in an authorship attribution problem.

Rule R1: P → c is said to be specific with regard to rule R2: P' → c', if and only if P ⊇ P'.

While in general, this pruning is harmless to accuracy and effective at making the process more efficient, one of the contributions is to prioritize more specific rules rather than more general ones. CMARAA, Classification by Multiple Association Rule for Authorship Attribution, orders rules slightly differently, changing condition 3) above to:

3) conf(R1) = conf(R2) and sup(R1) = sup(R2) but |ant(R1)| > |ant(R2)|

Part of the first round of pruning in CMARAA is, therefore, the opposite of what is done in CMAR [162]. More specific rules with a higher ranking are selected over more ambiguous rules. Intuitively the writing style patterns of an author should be as precise as possible to accurately represent the frequently occurring textual measurements. This change, in concert with other contributions that define CMARAA, allows the algorithm to achieve better results in terms of classification accuracy.

More general and more specific rule pruning is pursued when the rule is first inserted into the Classification Rule (CR) tree. When this happens, to check if the rule can be pruned or if other already inserted rules can be pruned, a retrieval over the tree is triggered.

The second round of pruning is done by only using rules that are positively correlated. Chi-square testing has been used to determine whether P is positively correlated with c for each rule. Only the positively correlated rules, i.e., those that pass a user-supplied significance level threshold, are used during classification. All rules that fail the correlation test are pruned.

Chi-square correlation-based pruning is used to reflect only strong implications to perform classification. By removing rules that are not positively correlated, noise is reduced, and this makes the classification process more efficient without negatively affecting accuracy. This pruning step is performed when a rule is being inserted into the CR-tree since the values necessary for performing the chi-square test are readily available at this point.

Algorithm 9.1 Database Coverage Rule-Based Selection
Input: a list of rules and a database coverage threshold T

1. **Output**: a concise but representative subset of class association rules

   Protocol

1. Order rules by rank:
2. For each training record, set the cover-count to zero:
3. For each rule R, find all matching training records.

   If rule R can appropriately classify at least one record, increase the cover-count of all records matching rule R by one. Remove a training record once its cover-count passes the database coverage threshold T.

   The third pruning method builds a subset of high-quality classification rules by performing database coverage pruning. A database coverage threshold [167] is used to reduce the number of CAR's significantly while maintaining the same representative number of rules per training record. This process is described in Algorithm 9.1.

   The database coverage method used by Li et al. [162] is similar to the one used by Liu et al. [160]. The primary difference is that, instead of removing one data object from the training dataset immediately after it is covered by some selected rule, it is left as part of the training set until such time that it is covered by a minimum of three other rules. The effect of this difference is that there will be more rules to consult when attributing a new object and therefore the unknown object will have a better chance of being classified accurately. This pruning is performed once the rule mining process has been completed; it is the last pruning of rules described in CMAR [162].

   One of this chapter's contributions is the addition of another round of pruning for CMARAA. This last pruning method has been brought over from AM [64, 149] and is called common frequent item set elimination. When rules are being inserted into the CR-tree, any rule with the same antecedent as another distinct rule is flagged for removal. Once the CR-tree is processed, the flagged rules are removed. The reason that common rules are not removed immediately once discovered is that another rule for another author that is also common may also exist. It is, therefore, necessary to leave all rules in place until the process of generating all CAR's is complete.

9.2.3　Authorship Classification

Once a set of rules is discovered and pruned for classification, as discussed in Sects. 9.2.1 and 9.2.2, the system is ready to classify anonymous e-mails. Given a test record, the subset of matching rules from the CAR list is selected. The rest of this section outlines how best to assign a class label based on the generated subset of rules.

   If all rules that match the target object share the same class label, the test record is associated with an author without contest.

   If two or more rules exist with different class labels, groups of rules are created for each class. All grouped rules share the same class label and each group is associated with a unique label. Then, the strength of each group is compared, and a record is associated with the strongest one.

To appropriately compare the groups' strengths, the combined effect of each group should be measured. Intuitively, if the group's rules are highly positively correlated and have a relatively high support metric, the group's effect should be strong.

Typical AC algorithms use the strongest rule as a representative, which means that the single rule with the highest rank is selected. The danger of simply choosing the rule with the highest rank is that this may favor minority classes, as illustrated by Example 9.2.

Example 9.2 In an authorship attribution exercise, there are two rules:

R1: FeatureA = no → AuthorB(support = 450, confidence = 60%)
R2: FeatureB = yes → AuthorA(support = 200, confidence = 99.5%)

See observed and expected values for rules R1 and R2 in Table 9.2.

Based on the measured and expected values, the chi-square value is 97.6 for R1 and 36.5 for R2. For an anonymous e-mail with no feature A and feature B, we may predict that the author would be Author B rule R1 if the choice between rules is based only on chi-square values. However, rule R2 is clearly much better than rule R1 since rule R2 has much higher confidence. This presents a challenge in determining which rule is the strongest. ∎

Table 9.2 Observed and expected values

(a) R1 observed			
RI	Author A	Author B	Total
Feature A	410	40	450
No feature A	20	30	50
Total	430	70	500

(b) R2 observed			
R2	Author A	Author B	Total
Feature B	209	1	210
No feature B	241	49	290
Total	450	50	500

(c) R1 expected			
R1	Author A	Author B	Total
Feature A	387	63	450
No feature A	43	7	50
Total	430	70	500

(d) R2 expected			
R2	Author A	Author B	Total
Feature B	189	21	210
No feature B	261	29	290
Total	450	50	500

Using the compound of correlation of rules as a measure is one alternative. For example, we can sum up the values in a group as the strongest measure of the group, but this would present the same problem in that it may favor minority classes.

A better way would be to integrate both correlation and popularity into the group measure and so a weighted measure [167], called Weighted Chi-Square (WCS), has been adopted. For each rule R: P → c, let sup(c) be the number of records in the training data set that are associated with class label c and let |T| be the number of data records in the entire training data set. Equation 9.1 defines the max Chi-square value, used as the upper bound of the Chi-square value of the rule.

$$\max X^2 = \left(\min\big(\sup(P),\sup(c)\big) - \frac{\sup(P)\sup(c)}{|T|} \right)^2 |T| e \qquad (9.1)$$

where

$$e = \frac{\dfrac{1}{\sup(P)\sup(c)} + \dfrac{1}{\sup(P)\big(|T|-\sup(c)\big)} + \dfrac{1}{\big(|T|-\sup(P)\sup(c)\big)} +}{\dfrac{1}{\big(|T|-\sup(P)\big)\big(|T|-\sup(c)\big)}}$$

For each group of rules, the weighted measure of the group is calculated using Eq. 9.2.

$$\sum \frac{x^{2^2}}{\max X^2} \qquad (9.2)$$

As demonstrated, the ratio of the X^2 value against its upper bound, max chi-square, is used to overcome the bias of the Chi-square value favoring any minority class. Liu et al. [162] noted that it was difficult to theoretically prove the soundness of measures based on the strength of rule groups. Instead, they tested and assessed the effect of measures through empirical observation and according to their experimental results, WCS got the best results in comparison to other sets of candidates.

9.3   Experimental Evaluation

To evaluate the accuracy, efficiency, and scalability of the Classification using the Multiple Association Rule (CMAR) [162] algorithm and the proposed augmented implementation of it, CMAR for Authorship Attribution (CMARAA), the result from a comprehensive performance study has been compiled. In this section, CMAR [162] and CMARAA have been compared against two well-known classification methods: Classification by Association (CBA) [160] for comparison against a baseline, AC algorithm, and AuthorMiner [64, 149] (AM), the previous leader in data mining-based authorship attribution. In addition, CMARAA is evaluated against some well-known classifiers that are commonly used in most authorship studies including Naive Bayes, BayesNet, Ensemble of Nested Dichotomies (END), and Decision Trees (e.g., J48). It has been shown that CMARAA generally outperforms the other methods including CBA [30] and AM [64, 149] when it comes to average accuracy.

All tests were performed on a 3.4GHz Core i7 with 12G main memory, running Mac OS 10.7.3. CMAR [162] and CBA [160], were implemented by Frans Coenen, in the course of demonstrating the power and scalability of their Apriori-TFP method [164]. AuthorMiner [64, 149] was implemented by its authors.

The e-mail collections used in this study are all from the publicly available Enron e-mail data set.[1] Specifically, hundreds of e-mails were manually selected from authors Paul Allen, John Arnold, Sally Beck, John Dasovich, Mark Headicke, Vince Kaminsky, Steven Kean, Kam Keiser, Philip Love, Kay Mann, Susan Scott, Carol St Clair, Kate Symes, and Kim Watson. E-mail headers were removed using a bash script and all content not written by the respective authors for the specific message was removed manually. The purpose of this cleaning was to ensure that each e-mail consisted of text written solely by its author. In general, only about 20% of e-mails in the original data set were retained as the rest consisted of text shorter than a few sentences, forwarded materials, or attachments. Similarly, for test objects, it would not be prudent to expect the accurate classification of any message consisting of a few words or a single sentence. It would be unhelpful to alter messages that short since they contain no specific writing style whatsoever, or even a spoofed writing style if the malicious entity was well versed in authorship attribution methodologies.

All results are accuracy averages compiled from running the various algorithms on data sets with the same number of authors, but each contains a different combination of authors, in order to show that results are stable and not simply the result of lucky guesses or hand-picked sets that support the conclusions. As expected, sets of authors with low accuracies show lower accuracies across the board and vice-versa. Each test is fair, and the data sets were not altered in any way, so no advantage was given to any-one algorithm over the others. For the experiments, the parameters are set as follows.

For CBA and CMAR, the support threshold is set to 10%, and the confidence threshold is set to 0%. The reason the confidence threshold has been set to 0% is to demonstrate the effect the final round of pruning in CMARAA has on accuracy. A high confidence threshold would effectively eliminate common rules among multiple authors as explained previously. Furthermore, setting a different confidence threshold for the various algorithms would make fair comparison impossible. Given that the confidence threshold in CMAR, CBA, and CMARAA does not affect accuracy, it was deemed safe and fair to simply set it to 0%. Please note that the confidence and support thresholds are set mainly to improve efficiency, not accuracy.

CMARAA uses the same support and confidence thresholds as CBA and CMAR but the support threshold is considered on a per author basis instead of applying it

---

[1] http://www.cs.cmu.edu/~enron/

to the size of the entire training data set. This allows for better extraction of frequent patterns on a per author basis. This is important because, without this distinction, CMARAA would technically be looking for frequent patterns across all authors instead of patterns that are representative for a single author. Without this contribution, it would be harder to associate anonymous e-mails with their author, as the results demonstrate quite clearly.

All figures and results consider the percentage of correctly matched authors in the testing set to reflect the accuracy of authorship identification. For example, if there are 4 test records and 3 of them are correctly matched to an author, the accuracy will be 75%. The tests have been performed by splitting the entire dataset into training and testing partitions with a ratio of 90% training data to 10% testing data. This means that if there are 1000 records in the entire data set, 900 records will be used for training and 100 for testing. For AuthorMiner and the AC algorithms, the training sets are used to discover frequent patterns for the classifier and then each e-mail in the test set is classified and verified. The training and test set splits are done on a per author basis, so each author's data set is split by the user-supplied percentage and the respective author sets are combined into a global training set and testing set. This separation is performed using the same method for all tested algorithms for results to be directly comparable.

In order for tests to be repeatable, the training and testing set split is done in order and not at random, with the first portion belonging to the training set and the rest to the test set. E-mails are named numerically and are input in ascending alphabetical order. This does potentially cause the results to be skewed towards how easy or hard it is to classify e-mails found at the end of the set, but all algorithms are forced to deal with this issue, again ensuring that results are comparable and fair to the strengths of each algorithm. The rest of this section describes the figures showing the results of the various tests.

Figure 9.2 shows the average classification accuracies over sets of collections of e-mails for 2–10 authors for each algorithm tested in this study. Accuracies range

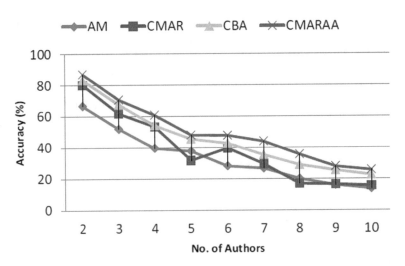

Fig. 9.2 Accuracy vs. number of authors [166]

from 30 to 92% with the most accurate algorithms being CMARAA, CBA, Classification by CMAR, and AuthorMiner (AM), respectively. The accuracies decrease as the number of authors increases, with CMAR seeing the biggest drop when there are ten authors. The classification result depends more on how unique the writing styles of the included authors are than the strength of the algorithms; naturally, two authors who have similar writing styles will be harder to tell apart, regardless of which technique is employed. With the minimum support threshold held constant, CMAR no longer generates rules with every run when there are many authors; this is due to the fact that the writing styles of authors are usually distinct from one another and so the support of feature items will not always pass the minimum support threshold, which is a percentage of records across the entire training set. For example, if the support threshold is 10% and there are ten authors with 100 e-mails each, then a feature item unique to one author would need to appear in every single e-mail in order to be considered frequent. This characteristic hurts CMAR as the number of authors increases. One of the main contributions of this study, implemented in CMARAA, reduces the support threshold to consider only one author's training set, allowing feature items that are frequent, even if unique to one author, to be considered.

Finally, the proposed method is evaluated against the classifiers that are commonly used in most authorship studies. Figure 9.3 depicts experimental results of AuthorMiner, Naive Bayes, Bayesian Networks (BayesNet), Classification by CMAR, CBA, END, Decision Trees (e.g., J48) and CMARAA. Showing the same trend as shown in Fig. 9.2, CMARAA also outperforms all other methods in this evaluation.

This study does not contain extensive CPU or I/O metrics. In the context of a criminal investigation, it is assumed that execution time is of little importance compared to the quality of classification and evidence collection. With that said,

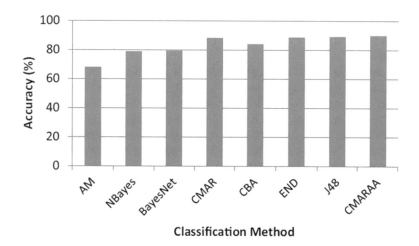

Fig. 9.3 Accuracy vs. classification method [166]

Table 9.3 Algorithm run time

Number of authors	Average algorithm run time (seconds)			
	AM	CMAR	CBA	CMARAA
2	15.717	3.132	3.0599	8.1525
3	52.83	7.27	7.83	55.39
4	575.74	8.44	8.51	96.34
5	443.467	13.24	12.9844	108.7104

Table 9.3 shows the average number of seconds each algorithm took to complete the classification process on sets of 2, 3, 4, and 5 authors. The algorithms that required the least CPU time for two authors were CBA, CMAR, CMARAA and AM respectively. For three authors, CMAR came in first, with CBA, CMARAA and AM in second, third, and fourth place respectively. The same order of fastest times held true for four authors. For five authors, CBA came in first place followed by CMAR, then CMARAA, and finally AuthorMiner.

The run times presented in Table 9.3, are primarily the result of their respective frequent itemset discovery processes. AuthorMiner uses the original Apriori [139] algorithm for discovering frequent itemsets, whereas CMAR, CBA, and CMARAA all use a much faster variant of the FP-Growth [165] algorithm, using more efficient tree structures to represent frequent itemsets and class association rules. Technically AM could use the FP-Growth [165] method, but improving its performance was not a goal of this study.

These results demonstrate reliable and repeatable proof that authorship attribution machine learning methods can be very useful. This also proves, once again, that the writing styles of authors can be modeled by extracting patterns from transformed semantic content to create individually recognizable Writeprints.

Cybercrime investigations need a state-of-the-art computer-aided Writeprint modeling algorithm that can provide reliable evidence to support authorship attribution. When someone's innocence or guilt is on the line, it is very important to have access to the most accurate and efficient methods of linking suspects to the evidence collected.

This study has explored the application of a promising data mining technique called associative classification to the e-mail authorship attribution problem for the first time. Additionally, it has been proposed that class-based associative classification allows for the extraction of a more accurate and complete Writeprint. It is also acknowledged that modifying the rule pruning and ranking system described in the popular Classification by CMAR algorithm to prioritize more specific patterns can also provide a more accurate Writeprint. The removal of patterns common among various authors results in a relatively unique Writeprint that makes for more convincing evidence in a court of law. The presented customized associative classification method helps fight cybercrime by addressing the e-mail authorship attribution problem. It has not been claimed that the findings of the proposed method will be enough to convict a suspect, but they can certainly complement other evidence, and therefore positively influence the court's decision.

Chapter 10
Criminal Information Mining

In the previous chapters, the different aspects of the authorship analysis problem were discussed. This chapter will propose a framework for extracting criminal information from the textual content of suspicious online messages. Archives of online messages, including chat logs, e-mails, web forums, and blogs, often contain an enormous amount of forensically relevant information about potential suspects and their illegitimate activities. Such information is usually found in either the header or body of an online document. The IP addresses, hostnames, sender and recipient addresses contained in the e-mail header, the user ID used in chats, and the screen names used in web-based communication help reveal information at the user or application level. For instance, information extracted from a suspicious e-mail corpus helps us to learn who the senders and recipients are, how often they communicate, and how many types of communities/cliques there are in a dataset. Such information also gives us an insight into the inter and intra-community patterns of communication. A clique or a community is a group of users who have an online communication link between them. Header content or user-level information is easy to extract and straightforward to use for the purposes of investigation.

The focus of this chapter is to analyze the content or body of online messages for the purpose of extracting social networks and the users' topic of discussion. In this context, the problem is defined as follows: When given a suspect machine which has been confiscated from a crime scene, an investigator is asked to identify potential suspects who are associated with the primary suspect S. The task is to analyze the content of online documents exchanged between these suspects. This chapter is specifically focused on analyzing chat logs. The investigator is provided with a taxonomy of certain street terms, representing certain crimes that are generally found in cybercrime-mediated textual conversation.

F. Iqbal et al., *Machine Learning for Authorship Attribution and Cyber Forensics*, International Series on Computer Entertainment and Media Technology, https://doi.org/10.1007/978-3-030-61675-5_10

Though some studies on the forensic analysis of online messages do exist, most of them focus on only one small aspect of the cybercrime investigation process. For instance, [129] focuses on mining chat logs for collecting sociolinguistic characteristics of potential authors of anonymous chat documents. The aim of [65, 116] is to develop a classification model for predicting the true author of an anonymous e-mail message. Alfonse and Manandhar [168] applied named entity recognition, a subtask of information extraction, to extract information such as the names of people, organizations, places, or other contact information from textual documents. Minkov et al. [76] developed a technique for extracting namedentity information from informal documents such as e-mails. Chau et al. [112] proposed criminal link analysis techniques and Xiang et al. [139] proposed crime data visualization techniques for the Web and Internet-level communication of cybercriminals. In [91], a data mining framework is developed for analyzing different kinds of crimes.

In contrast, in the study we will look at in this chapter, a framework is presented for extracting and reporting forensically relevant information from malicious online textual communication documents. A highly important aspect of this framework is that the entire process is automated, including the retrieval of documents, the extraction of different kinds of information, and the intuitive presentation of the findings. The proposed framework consists of three modules including clique miner, concept miner, and information visualizer, as depicted in Fig. 10.1. A clique defined in this context is a group of entities co-occurring together in the textual contents of online messages. The clique miner is designed to first identify the named entities appearing in the given suspicious chat logs; it then groups them according to the frequency of their co-occurrence. For the identification module, Stanford Named Entity Recognizer is used, while the grouping phase is accomplished by employing the concept of frequent pattern mining [139]. Once the cliques are extracted, the concept miner retrieves the documents from each clique and extracts key concepts that reflect the theme of communication between members of that clique. The output of concept miner is a list of important terms (keywords), common concepts, key concepts, and a summary.

The information visualizer is used to objectively display the identified groups and the extracted information (e.g., keywords and concepts) by employing social networking concepts. In the visualized social network, depicted in Fig. 10.1, the nodes represent the entities while the arcs connecting the nodes indicate the exis-

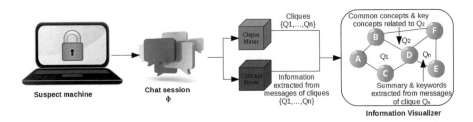

Fig. 10.1 Framework overview

tence of a relationship between the entities. The nodes and the arcs belonging to the same clique are labeled with the letter Q; the subscript i indicates the clique number. The cliques are labeled with the chat summary, keywords, common concepts, and key concepts extracted from the chat sessions of a specific group. The contributions of this study are listed below:

- **Analyzing unstructured data:** Most previous machine learning methods for criminal investigation focus on structured data, e.g., criminal police records. This data-mining framework is designed for analyzing online messages including chat logs.
- **Identifying topics dynamically:** Most topic identification methods assume to have some predefined topic categories with example documents. This approach does not need any training data. It can be employed to dynamically assign topics to unknown online messages based solely on the content of the documents in question.
- **Adapting domain knowledge:** By employing the presented approach, the investigator can incorporate domain-specific terms to obtain results that are more specific.

10.1 Problem Statement

Suppose an investigator has seized a computer from a suspect S. Let ϕ be the chat log obtained from some commonly used instant messaging systems, such as Skype, Yahoo! Messenger, or IRC, found by examining the content of the computer. Typically, a chat log consists of a set of chat sessions, where each chat session contains a set of text messages exchanged between suspect S and the chat users who appear in the friend list of S. The problem of criminal clique mining is to discover the communities (i.e., cliques) actively involved by the suspect S in ϕ, to identify the relationships among the members in the cliques, and to extract the concepts/topics that bring the cliques together. It is divided into two sub-problems: clique mining and concept analysis.

10.1.1 Subproblem: Clique Mining

The subproblem of clique mining is to efficiently identify all the cliques from a given chat log. The following observations about cliques were formulated after an extensive discussion with the digital forensic team of a Canadian law enforcement unit. An entity generally refers to the name of a person, a company, or an object identified in a chat log. For ease of discussion, it is assumed that an entity refers to a person's name in the rest of the chapter.

A group of entities is considered to be a clique in a chat log if they chat with each other frequently, or if their names appear together frequently in some minimum number of chat sessions.

This notion of a clique is more general than simply counting the number of messages sent between two chat users. An entity ε is considered to be in a clique as long as his/her name frequently appears in the chat sessions together with some group of chat users, even if ε has never chatted with the other members in the clique or even if ε is not a chat user in the log. Capturing such a generalized notion of a clique is important for real-life investigations because the members of a clique are not limited to be the chat users found in the log. Such a generalized notion often leads to new clues for further investigation. For example, two suspected entities ε_1 and ε_3 frequently mention the name of a third person ε in the chat because ε is their "boss" behind the scene. Thus, all three of them form a clique although ε may not be a user found in the chat log. Nonetheless, defining a clique in this relaxed way may increase the chance of identifying false positive cliques. For example, two suspects may frequently discuss EJ who is a celebrity and thus would not be considered part of the clique. Despite this, in the context of crime investigation, an investigator would rather spend time filtering out false positives than risk missing any potentially useful evidence.

A chat log ϕ is a collection of chat sessions $\{\phi_1 \cdots \phi_p\}$. Let $E(\phi) = \{\varepsilon_1, \cdots, \varepsilon_u\}$ denote the universe of all entities identified in ϕ. Let $E(\phi_i)$ denote the set of entities identified in a chat session ϕ_i, where $E(\phi_i) \subseteq E(\phi)$. For example, $E(\phi_5) = \{\varepsilon_4, \varepsilon_5, \varepsilon_7\}$ in Table 10.1. Let $Y \subseteq E(\phi)$ be a set of entities called entity set. A session ϕ_i contains an entityset Y if $Y \subseteq E(\phi_i)$. An entityset that contains k entities is called a k-entityset. For example, the entityset $Y = \{\varepsilon_4, \varepsilon_5, \varepsilon_7\}$ is a three-entityset. The support of an entity set Y is the percentage of chat sessions in ϕ that contain Y.

An entity set Y is a clique in ϕ if the support of Y is greater than or equal to some user-specified minimum support threshold.

Chat session	Identified entities
ϕ_1	$\{\varepsilon_2, \varepsilon_5, \varepsilon_7, \varepsilon_9\}$
ϕ_2	$\{\varepsilon_2, \varepsilon_5, \varepsilon_7\}$
ϕ_3	$\{\varepsilon_2, \varepsilon_5\}$
ϕ_4	$\{\varepsilon_1, \varepsilon_5, \varepsilon_7\}$
ϕ_5	$\{\varepsilon_4, \varepsilon_5, \varepsilon_7\}$
ϕ_6	$\{\varepsilon_3, \varepsilon_6, \varepsilon_8\}$
ϕ_7	$\{\varepsilon_4, \varepsilon_5, \varepsilon_8\}$
ϕ_8	$\{\varepsilon_3, \varepsilon_6, \varepsilon_8\}$
ϕ_9	$\{\varepsilon_2, \varepsilon_5, \varepsilon_8\}$
ϕ_{10}	$\{\varepsilon_1, \varepsilon_5, \varepsilon_7, \varepsilon_8, \varepsilon_9\}$

Table 10.1 Vector of entities representing chat sessions

Definition 10.1 (Clique). Let ϕ be a collection of chat sessions. Let support(Y) be the percentage of sessions in ϕ that contain an entityset Y, where $Y \subseteq E(\phi)$. An entityset Y is a clique in ϕ if support(Y) $\geq$ min_sup, where the minimum support threshold min_sup is a real number in an interval of [0, 1]. A clique containing k entities is called a k-clique.

Example 10.1 Consider Table 10.1. Suppose the user-specified threshold min_sup = 0.3, which means that an entityset Y is a clique if at least 3 out of the 10 sessions contain all entities in Y. Similarly, $\{\mathcal{E}_4, \mathcal{E}_5\}$ is not a clique because it has support 2/10 = 0.2. $\{\mathcal{E}_2, \mathcal{E}_5\}$ is a 2-clique because it has support for 4/10 = 0.4 and contains 2 entities. Likewise, $\{\mathcal{E}_5, \mathcal{E}_8\}$ is a 2-clique with support 3/10 = 0.3.

Definition 10.2 (Clique mining). Let ϕ be a collection of chat sessions. Let min_sup be a user-specified minimum support threshold. The subproblem of clique mining is to efficiently identify all cliques in ϕ with respect to *min_sup*.

10.1.2   Subproblem: Concept Analysis

According to the discussions with the Canadian law enforcement unit, they encountered some cases that involved thousands of chat users in the Windows Live Messenger chat log on a single machine. Consequently, there could be hundreds of cliques discovered in the chat log. The discovered cliques reflect the suspect's various social circles, including individuals from his/her family, friendship groups, work, and religion. To identify the cliques that are related to the suspect's criminal activities, the investigator has to analyze the content of the chat sessions of each clique. The subproblem of concept analysis is to extract the concepts that reflect the semantics of the conversations, not just a collection of keywords. To facilitate the process of concept analysis, it is assumed that a lexical database exists that is able to capture the conceptual hierarchies of a language, e.g., WordNet for English.

Definition 10.3 (Concept analysis), Let Q be a set of cliques discovered in ϕ according to Definition 10.2. Let $\phi(Qi) \subseteq \phi$ be the set of chat sessions contributing to the support of a clique $Q_i \in Q$. Note that the same chat session may contribute to multiple cliques. Let H be a lexical database of the same language used in ϕ. The subproblem of concept analysis is to extract a set of key concepts, denoted by $KC(Q_i)$, for each discovered clique $Q_i \in Q$ by using the lexical database H. The key concepts represent the topics that bring the group of entities together to form a clique.

10.2   Proposed Approach

Figure 10.2 depicts an overview of the proposed framework, which consists of three components including clique miner, concept miner, and information visualizer. Clique miner identifies all the cliques and their support from the given chat log. Concept miner analyzes the chat sessions of each identified clique and extracts the key concepts of the conversations. Information visualizer provides a graphical interface to allow the user to interactively browse cliques at different abstraction levels. Each module is described separately in the following paragraphs.

10.2.1   Clique Miner

The process of clique mining consists of three steps:

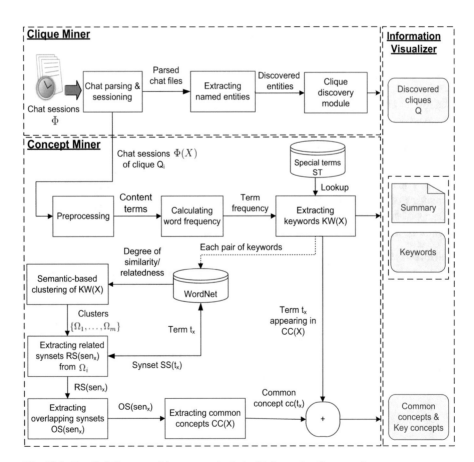

Fig. 10.2 Detailed diagram of the proposed criminal information framework

1. Dividing chat log into sessions: A session is a sequence of messages exchanged between a group of chat users within a "logical" period. For instance, in Windows Live Messenger, a session with a person P begins when the first message is sent between P and the suspect S and ends when the suspect closes the chat log window with P. Once the chat log window is closed, the re-initiation of the chat is considered a new session with a new session ID in the log. When it comes to the IRC log on a public chat room, the situation is more complicated because multiple users can chat simultaneously and there are no logical breakpoints for breaking a log into sessions. A simple solution is to break down the log into sessions by some predefined unit of time, say by 15 min. Further to this, a better solution is to look for the time gap between messages and consider a new session when the time gap is larger than a short period of a certain duration, for example, 1 min.

2. Extracting entities: Next, the existing Named Entity Recognition (NER) tools are employed to extract entity names from each chat session, e.g., person, location, time, etc. In this study, it is assumed an entity is a person, but in a real-life application, an entity can also be an organization, a location, a phone number, or a website [91]. NER systems use linguistic grammar-based techniques and statistical models. Handcrafted grammar-based systems typically obtain better results, but at the cost of months of work by experienced computational linguists. Statistical NER systems typically require a large amount of manually annotated training data. In this study, Stanford Named Entity Recognizer2 software called CRFClassifier was used. The software is based on linear chain Conditional Random Field (CRF) sequence models. It is trained on widely used named entity corpora. Other NER tools can be employed if the document files contain non-English names, as NER is not the focus in this chapter. The next step, clique mining, operates on a data table consisting of records of entities that represent entities in session, not on the actual chat log. (3) Mining cliques: Recall that an entityset Y is any combination of entities identified in the chat log. An entityset is a clique if its support is equal to or greater than a given threshold. A naive approach is to enumerate all possible entitysets and identify the cliques by counting the support of each entityset in ϕ. However, if the number of identified entities $|E(\phi)|$ is large, it is infeasible to enumerate all possible entitysets because there are $2^{|E(\phi)|}$ possible combinations. The Apriori algorithm [139], which was originally designed to extract frequent patterns from transaction data, has been modified to efficiently extract all cliques from ϕ. the modified algorithm is described as follows.

Recall that $E(\phi)$ denotes the universe of all entities in ϕ, and $E(\phi_i)$ denotes the set of entities in a session, $\phi_i \in \phi$, where $E(\phi_i) \subseteq E(\phi)$. The proposed Clique Miner (CM) is a level-wise iterative search algorithm that uses the k-cliques to explore the (k + 1)-cliques. The generation of (k + 1)-cliques from k-cliques is based on the following CM property.

Property 10.1 (CM property). All nonempty subsets of a clique are also cliques because support(Y') ≥ support(Y) if Y' ⊆ Y.

By definition, an entityset Y is not a clique if support(Y) < min_sup. The above property implies that adding an entity to an entityset that is not a clique will never make the entityset become a clique. Thus, if a k-entityset Y is not an entityset, then there is no need to generate (k + 1)-entityset Y U {ε} because Y U {ε} must not be a clique. The closeness among the entities in a clique Y is indicated by |φ(Y)|, which is the support of Y. CM can identify all cliques by efficiently pruning the entitysets that are not cliques based on the CM property.

Algorithm 10.1 summarizes the proposed Clique Mining Algorithm (CM). The algorithm identifies the k-cliques from the (k − 1)-cliques based on the CM property. The first step is to find the set of 1-cliques, denoted by Q_1. This is achieved by scanning the chat log data table once and calculating the support count for each 1-clique. Q_1 contains all 1-cliques X with support(C_j) ≥ min_sup. The set of 1-cliques is then used to identify the set of candidate 2-cliques, denoted by Candidates$_2$. Then the algorithm scans the table once to count the support of each candidate X in Candidates$_2$. All candidates X that satisfy |φ(X)| ≥ min_sup (i.e., which have support greater than or equal to a threshold) are 2-cliques, denoted by Q_2. The algorithm repeats the process of generating Q_k from Q_{k-1} and stops if Candidate$_k$ is empty.

Algorithm 10.1 Clique Mining Algorithm

```
Input:  Chat log Φ
Input:  Minimum support threshold min_sup
Output: Cliques Q = {Q₁ U···U Qₖ}
Output: Chat sessions Φ(X), ∀X ∈ Q
 1:  Q₁ ← {ε | ε ∈ E(Φ)∧ support({ε}) ≥ min_sup};
 2:  for (k = 2; Qₖ₋₁ ≠ 0; k++) do
 3:      Candidatesₖ ← Qₖ₋₁ ⋈ Qₖ₋₁;
 4:      for all entityset Y ∈ Candidatesₖ do
 5:          if ∃Y' ⊂ Y such that Y' ∉ Qₖ₋₁ then
 6:              Candidatesₖ ← Candidatesₖ − Y;
 7:          end if
 8:      end for
 9:      Φ(X) ← 0, ∀X ∈ Candidatesₖ;
10:      for all chat session φ ∈ Φ do
11:          for all entityset X ∈ Candidatesₖ do
12:              if X ⊆ E(φ) then
13:                  Φ(X) ← Φ(X)∪φ;
14:              end if
15:          end for
16:      end for
17:      Qₖ ← {X | X ∈ Candidatesₖ∧|Φ(X)| ≥ min_sup};
18:  end for
19:  Q = {Q₁U···UQₖ};
20:  return Q and Φ(X), ∀X ∈ Q;
```

Lines 9–17 describe the procedure of scanning the data table and keeping track of the associated document of each clique X in Candidates$_k$. Each candidate entity-set X is looked up in the entities of each chat session E(ϕ). If a match is found, the chat session ϕ is added to the set ϕ (X). If the support |ϕ (X)| is greater than or equal to the user-Specified minimum threshold min_sup, then X is added to Q$_k$, the set of k-cliques with k members. The algorithm terminates when no more candidates can be generated or when none of the candidate entitysets passes the min_sup threshold. The algorithm returns all cliques Q = {Q$_1$ $\cup\cdots\cup$ Q$_k$}, except for the 1-cliques, with their associated chat sessions.

The following example shows how to efficiently extract all frequent patterns.

Example 10.2 Consider Table 10.1 with min_sup = 0.3. First, identify all the enti-ties by scanning the table once to obtain the support of every entity. The entities having support $\geq$0.3 are 1-cliques Q$_1$ = {{$\mathcal{E}_2$}, {$\mathcal{E}_5$}, {$\mathcal{E}_7$}, {$\mathcal{E}_8$}}. Then join Q$_1$ with itself, i.e., Q$_1$ $\bowtie$Q$_1$, to generate the candidate set Candidates$_2$ = {{$\mathcal{E}_2$, $\mathcal{E}_5$}, {$\mathcal{E}_2$, $\mathcal{E}_7$}, {$\mathcal{E}_2$, $\mathcal{E}_8$}, {$\mathcal{E}_5$, $\mathcal{E}_7$}, {$\mathcal{E}_5$, $\mathcal{E}_8$}, {$\mathcal{E}_7$, $\mathcal{E}_8$}} and scan the table once to obtain the support of every entityset in Candidates$_2$. Next, identify the 2-cliques Q$_2$ = {{$\mathcal{E}_2$, $\mathcal{E}_5$}, {$\mathcal{E}_5$, $\mathcal{E}_7$}, {$\mathcal{E}_5$, $\mathcal{E}_8$}}. Next, perform Q$_2$ $\bowtie$ Q$_2$ to generate Candidates$_3$ = {$\mathcal{E}_2$, $\mathcal{E}_5$, $\mathcal{E}_7$} and determine Q3 = $\emptyset$. Finally, the algorithm returns Q$_2$ and the associated chat sessions of every clique in Q$_2$.

10.2.2   Concept Miner

The purpose of this phase is to analyze the chat sessions and summarize the content into some high-level topics to facilitate effective browsing in the visualization phase. The concept of miner extracts the underlying semantics of the written words from the set of associated chat sessions ϕ (X) of every clique X $\in$ Q identified by Algorithm 10.1. It is important to identify the underlying semantics of the written words as many perpetrators use different obfuscation and deception techniques to covertly conduct their illegitimate activities. Understanding the semantic and con-textual meaning of online messages is difficult because they are unstructured and are usually written in paralanguage. The abbreviations, special symbols, and visual metaphors used in malicious messages convey special meanings and are meaningful in some specific context. Specifically, the concept miner extracts three notions from ϕ (X): Keywords are frequent words extracted from ϕ(X). Common concepts are high-level topics shared by the chat sessions in ϕ(X). Key concepts are the top-ranked concepts by importance.

Algorithm 10.2 Concept Mining Algorithm

Input: Cliques Q from Algorithm 8.3
Input: Associated chat sessions $\Phi(X), \forall X \in Q$
Input: Search terms ST
Input: Keyword threshold α
Input: Maximum number of key concepts β
Output: Keywords $KW(X), \forall X \in Q$
Output: Common concepts $CC(X), \forall X \in Q$
Output: Key concepts $KC(X), \forall X \in Q$
Output: Miscellaneous information $MiscInfo(X), \forall X \in Q$

1: **for all** $X \in Q$ **do**
2:     $KW(X) \leftarrow \{t \mid t \in \Phi(X) \wedge t \in ST$ or t with top α $tf\_idf(t)\}$;
3:     Group the terms in $KW(X)$ into clusters $\{\Omega_1, \dots, \Omega_m\}$;
4:     $CC(X) \leftarrow \emptyset$;
5:     **for all** cluster $\Omega_i \in \{\Omega_1, \dots, \Omega_m\}$ **do**
6:       **for all** term $t_x \in \Omega_i$ **do**
7:         $SS(t_x) \leftarrow$ synsets of t_x from WordNet;
8:         **for all** sense $sen_x \in SS(t_x)$ **do**
9:           $RS(sen_x) \leftarrow$ related synsets of sen_x from WordNet;
10:          $OS(sen_x) \leftarrow RS(sen_x) \cap RS(sen_y), \forall sen_y \in SS(t_y)$, where $\forall t_y \in \Omega_i, t_x \neq t_y$;
11:         **end for**
12:         $CC(X) \leftarrow CC(X) \cup OS(BestSen)$, where $BestSen \in SS(t_x)$ is the sense having the largest $|OS(BestSen)|$;
13:       **end for**
14:     **end for**
15:     **for all** common concept $cc \in CC(X)$ **do**
16:       $Score(cc) \leftarrow 0$;
17:       **for all** term $t \in KW(X)$ **do**
18:         **if** $t \in cc$ **then**
19:           $Score(cc) \leftarrow Score(cc) + tf\_idf(t)$;
20:         **end if**
21:       **end for**
22:       $Score(cc) \leftarrow Score(cc)/|cc|$;
23:     **end for**
24:     $KC(X) \leftarrow \{cc \mid cc \in CC(X)$ with top β $Score(cc)\}$;
25:     $MiscInfo(X) \leftarrow$ various information identified in $\Phi(X)$;
26: **end for**
27: **return** $KW(X), CC(X), KC(X)$, and $MiscInfo(X), \forall X \in Q$;

Algorithm 4: Concept Mining Algorithm

Algorithm 10.2 provides an overview of the Concept Mining Algorithm.

For every clique $X \in Q$, the keywords from $\phi(X)$ are extracted and grouped into clusters $\{\Omega_1, \cdots, \Omega_m\}$ by semantics. Following this, the common concepts CC among the keywords within each cluster Ω are extracted, and finally the most important ones, the key concepts, are identified. These five steps are elaborated as follows. The first task is to apply some standard text mining preprocessing procedures to the input chat log ϕ. Such procedures are outlined here. Tokenization involves breaking a sentence into a sequence of words called terms. Stop word removal is applied to remove the context-independent words, which do not contribute to identifying the semantics of the text. Stop words include function words (e.g., 'is', 'my', 'yours', and 'below'), short words (e.g., words containing 1–3 characters), punctuation, and non-informative symbols and characters [63, 110]. Stemming involves converting different forms of the same word into the root word [108, 109]. For instance, the words compute, computed, computer, and computing are converted into the root word compute. After preprocessing, each chat session $X \in \phi$ is represented as a vector of terms [112].

1. **Extracting keywords (Line 2):** There are two kinds of keywords. A term t in $\phi(X)$ is a keyword of X, denoted by KW (X), if it appears in the list of user-specified special terms or if it occurs frequently in many chat sessions of one clique but not frequently in the chat sessions of other cliques:

   - Although may not appear frequently, some special terms are important for crime investigations. For instance, certain crime-relevant street terms such as marijuana, heroin, or opium are relevant and therefore such terms require more attention even though they may appear only once. To identify such special terms, the investigator can specify a list of special terms, denoted by ST. In this implementation, the terms are collected from different law enforcement agencies and online sources.
   - A term is important in $\phi(X)$ if it appears frequently in the chat session $\phi(X)$ of clique $X \in Q$ but not frequently in chat session ϕ (Y) of another clique Y $\in Q$, where $X \neq Y$. Intuitively, these terms can help differentiate the topic of a clique from others. To identify them, the tf − idf of every term is computed as discussed and the top a of them are added to KW (X), where α is a user-specified threshold. The sentences containing the keywords are key sentences that can be used for summary [169].

2. **Clustering keywords by semantics (Line 3):** The objective of this step is to group the keywords into clusters $\{\Omega_1, \cdots, \Omega_m\}$ such that the keywords in the same cluster have high similarity and the keywords in different clusters have low similarity. In the literature of natural language processing, semantic similarity is called paradigmatic similarity, and relatedness is known as syntagmatic similarity [170]. Two words are paradigmatically similar if they can be substituted by each other in a specific context without changing semantics of the sentence too much. For instance, the word price can be replaced by cost in the sentence "The price of the monitor is high". Two words are syntagmatically similar if they often appear together, for example, the words knife and cut often appear together. The agglomerative hierarchical clustering method has been employed to create the clusters [171]. The general idea is to compare every pair of terms in KW (X) and iteratively merge the pairs with the highest similarity. Similarity is measured by the semantic relatedness of word senses according to similarity software WordNet. Specifically, WordNet is employed to compute the paradigmatic and syntagmatic similarity. Please note that it is important to cluster the words by semantics first; otherwise, it will be difficult to find common concepts in the next step.

3. **Extracting common concepts (Lines 4–14):** Next, the objective is to identify some common concepts that cover the semantics of the keywords of each cluster $\Omega_i \in \{\Omega_1, \cdots, \Omega_m\}$ by making use of the WordNet. In WordNet, every term t is

associated with a set of senses called synset. Each sense contains a set of terms that represents the interpretation of the term t in a specific context. Consider Table 10.2 for example. The term coke has three senses (synsets). In the context of drug trafficking, coke means cocaine but in other contexts, it means a type of drink or carbon fuel. Below, we describe how to select the most suitable sense of each term based on the context described by other terms in the same cluster.

The following operations have been performed for every keyword t_x in each cluster Ω_i. First, the related synsets, denoted by RS (sen_x), are obtained including the synonyms, direikct hypernyms, and entailments for every sense sen_x of t_x. Second, the overlapping related synsets of sen_x and of every other term t_y in the same cluster Ω_i are identified. The overlapping synsets are denoted by OS(sen_x). Finally, the best sense is selected, denoted by BestSen, that has the largest number of overlapping synset, and add OS(BestSen) to the common concepts of clique X, denoted by CC(X). Table 10.2 lists the senses (synsets) of some terms followed by a direct hypernym of the sense. Suppose we find the keywords coke

Table 10.2 Synsets and direct hypemyms of selected terms retrieved from WordNet

Term	Synsets ⇒ Direct hypernyms
Snow	1. Snow, snowfall—(precipitation falling from clouds in the form of ice crystals) ⇒ Precipitation, downfall—(the falling to earth of any form of water (rain or snow or hail or sleet or mist)) 2. Snow—(a layer of snowflakes (white crystals of frozen water) covering the ground) ⇒ layer—(a relatively thin sheet like expanse or region lying over or under another) 3. Snow, C. P. Snow, Charles Percy Snow, Baron Snow of Leicester—(English writer of novels about moral dilemmas) => Writer, author—(writes (books or stories or articles or the like) professionally (for pay)) 4. Snow, coke, blow, nose candy, C—(street names for cocaine) ⇒ Cocaine, cocain—(a narcotic (alkaloid) extracted from coca leaves; used as a surface anesthetic or taken for pleasure)
Coke	1. Coke—(carbon fuel produced by distillation of coal) ⇒ Fuel—(a substance that can be consumed to produce energy; "more fuel is needed during the winter months") 2. Coke, Coca-Cola—(Coca Cola is a trademarked cola) ⇒ Cola, dope—(carbonated drink flavored with extract from kola nuts ('dope' is a southernism in the United States)) 3. Coke, blow, nose candy, snow, C—(street names for cocaine) ⇒ Cocaine, cocain—(a narcotic (alkaloid) extracted from coca leaves; used as a surface anesthetic or taken for pleasure)
Nose candy	1. Coke, blow, nose candy, snow, C—(street names for cocaine) ⇒ Cocaine, cocain–(a narcotic (alkaloid) extracted from coca leaves; used as a surface anesthetic or taken for pleasure)
Cocaine	1. Cocaine, cocain—(a narcotic (alkaloid) extracted from coca leaves; used as a surface anesthetic or taken for pleasure) ⇒ Hard drug—(a narcotic that is considered relatively strong and likely to cause addiction)

and snow in some of a certain clique's chat sessions. By intersecting their related synsets including the direct hypernyms, a common concept {coke, blow, nose candy, snow, C}, which has a direct hypernym {hard drug} can be identified. Without considering the terms coke and snow in the correct context, the terms will probably be misinterpreted.

4. **Identifying key concepts (Lines 15–24):** According to the evaluation, the semantics of the chat sessions associated with a clique are well captured by the common concepts extracted. However, in a real application, there are too many of them. It is impractical to display all the common concepts in the interactive user interface for browsing the cliques. Thus, the common concepts are ranked and the top ß of them are displayed, where ß is a user-specified threshold. Intuitively, a common concept is a key concept in X if its senses contain a keyword found in the clique X. The importance of a term is computed by the tf − idf value. The importance of a common concept is the sum of the tf − idf values of the matched terms normalized by the number of terms in the common concept.

5. **Extracting miscellaneous information (Line 25):** This step involves extracting relevant information, such as phone numbers, addresses, e-mails, and website URLs, from the chat sessions of every clique. This task can be easily achieved by matching with some regular expressions.

10.2.3 Information Visualizer

The objective of the information visualizer is to provide an interactive user interface to browse the discovered cliques and the relevant information. In general, a clique can be displayed as a graph, in which the nodes represent the entities, the edges represent the relationship, and the lengths of edges represent the closeness between the entities. However, the visualization task in this study is challenging when the number of discovered cliques is large. Recall that Property 10.1 states that every subset of a clique is also a clique, so the discovered cliques, in fact, represent multiple layers of relationships. Each clique has its own closeness, keywords, common concepts, key concepts, and other relevant information. An intuitive interface has been designed by integrating a data visualization tool, called prefuse [172], which allows the user to drill-down and roll-up on a clique. Prefuse is a collection of software tools, written in Java, and is used for creating interactive data visualization solutions. See the next section for more details.

10.3 Experiments and Discussion

There are four objectives in this section. (1) To verify if the cliques, extracted by the clique miner represent a meaningful group of individuals in the real world and to measure the effect of a minimum support threshold on the number of cliques.

(2) To evaluate whether or not the concept miner can precisely identify the important keywords, common concepts, and key concepts from the chat conversation of each extracted clique. (3) To quantitatively measure the efficiency of the developed framework in terms of the total execution time versus the user-defined minimum support threshold. (4) To measure the scalability of the presented framework by plotting the execution time vis-à-vis data size. Finding a real-life dataset for evaluating the proposed approach is not trivial due to privacy issues.

Most law enforcement agencies and private organizations have access to criminal data, but they cannot make it public due to legal constraints. For instance, the chat corpus, collected by Perverted Justice, is a rich source of predominantly cybercrime related data and is available online but it cannot be used for analysis without the consent of the people concerned. These chat logs mostly contain cyber predatory and cyberbullying conversations between predators and the pseudo-victims. Due to such privacy issues, the research team objectively creates MSN chat logs in which one of the team members pretends to be the primary suspect chatting with different users. In the given chat conversation, one of the members behaves like a pseudo-drug dealer by using the street names of some drugs in his conversations with the primary suspect.

In the first set of experiments, the clique miner takes the given chat log and displays the identified cliques. Figure 10.3, representing the screenshot of the framework, displays the graphical view of the discovered cliques. By using the *GUI* of the developed framework, the user can identify a clique by moving the mouse on the figure. The group of entities representing one clique is highlighted together. In the figure, ten cliques, each containing 2–3 entities are shown. The central node in each clique denotes the primary suspect and the peripheral nodes represent the entities

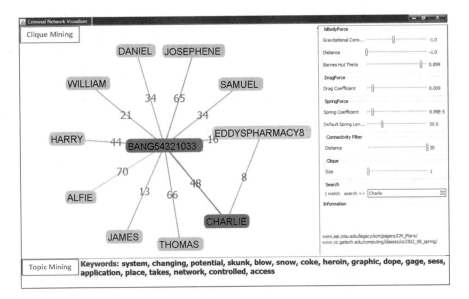

Fig. 10.3 A sample screenshot of the presented framework

associated with the suspect. The arcs connecting the entities indicate the existence of a relationship between the entities. The clique containing entities BANG54321033, EDDYSPHARMACY8, and CHARIE is interesting as the chat conversations between its members clearly contain drug-related terms, e.g., coke, dope and snow. The extracted entities and the discovered cliques were manually compared with the textual content of the given chat sessions. It was found that more than 80% of the cliques were correctly identified with few false positive cases. This success rate can be further improved by using sophisticated tools for named entity recognition. In the second set of experiments, the user-defined minimum support threshold was changed incrementally to check its effect on the number of cliques, as shown in Fig. 10.4. The number of cliques extracted from the given chat log spans from 155 to 8 for minimum support ranging from 0.33 to 3.33%. The number of cliques is inversely proportional to the minimum support, i.e., increasing the support will cause a decrease in the number of cliques. The number of cliques sharply drops by changing the support threshold from 0.33 to 0.66% for the chat log in question. The curve becomes almost flat when the support count is increased to 1.33%. There is always a tradeoff between the two parameters, and they can be adjusted according to the specific requirements of the investigation in question.

The third set of experiments is performed to evaluate the concept analysis functionality of the presented framework. The concept miner retrieves the chat log of each clique discovered in the clique-mining step and extracts the keywords, common concepts, and key concepts from each chat collection separately. Figure 10.3 visualizes the extracted cliques and the concept analysis results associated with each clique. The drill-down and roll-up capability of the framework allow the user to browse the cliques and the summaries of the conversations.

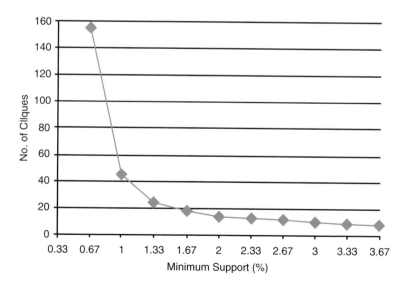

Fig. 10.4 Effect of minimum support on numbers of cliques

The results indicated that the concept analysis summary of the chat log belonging to the clique comprising BANG54321033, EDDYSPHARMACY8, and CHARLIE was particularly interesting. The extracted keywords included blow, snow, coke, dope, and gage, which are the street terms used to represent cocaine, a narcotic. The concept miner also identified a set of words, including system, changing, and potential, as keywords. This happened due to the high frequency of these words. The words 'cocaine' and 'cocain', identified as the key concepts, representing the topic of the chat conversation of the aforementioned clique. The other extracted information such as the message summary and the common concepts are not shown in the figure for simplicity. By comparing the extracted keywords and the related key concepts with the WordNet conceptual hierarchy (shown in Table 10.2), it is possible to conclude that the concept miner is successful at correctly identifying the topic of online messages.

The slide bars, denoted by NBodyForce in Fig. 10.3, are used by the user for setting the parameters. The user needs to specify the minimum support threshold and the size of the chat dataset. The fourth set of experiments is employed to measure the runtime efficiency of the proposed framework. For this, MSN chat logs with an initial size of 2.59 MB were used, voluntarily contributed by the team members. The value of total execution time (measured in seconds) is plotted against the minimum support, as shown in Fig. 10.5. The value of minimum support ranges from 0.33 to 3.33%. The total execution time is highest, i.e., 53 s for minimum support 0.66% and it decreases as the minimum support increases. The execution time drops sharply from 0.66 to 1.0% and remains flat afterward.

Generally, a software tool is considered scalable provided its execution time increases linearly as the size of input data increases. However, if the execution time grows exponentially with the increase in the data size, then the tool is not scalable.

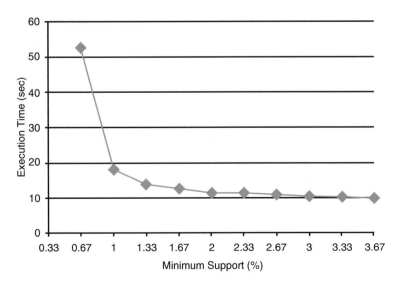

Fig. 10.5 Efficiency [Execution time vs. Minimum support]

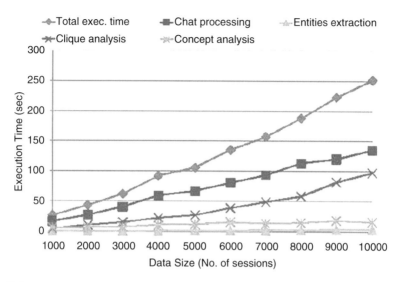

Fig. 10.6 Scalability [Execution time vs. Data size]

To measure the scalability, the size of the dataset (measured in terms of the total chat sessions) is incrementally changed while keeping the minimum support constant at 0.67% in clique miner. Initially, a dataset of 1000 sessions is used, which is incremented by an equal interval size of 2000 sessions up to a maximum size of 10,000 sessions. Depicted in Fig. 10.6, the execution times of each component of this framework were measured separately. Finally, all the individual scores were added together to obtain the total execution time of the entire framework code. From this graph, a linear increase in the execution time of each component as well as the sum of all the components can be clearly seen. The figure indicates that the proposed framework is scalable.

In this chapter, a criminal information-mining framework was presented and applied to the process of forensically extracting relevant information from suspicious online messages. The framework is designed to take online messages as input and provides a set of cliques and the topic of discussion of the chat conversation of each clique as output. The experimental result on a given chat log suggests that the proposed framework can precisely identify the pertinent cliques and the perceived meaning of the messages exchanged between the members of each clique. The framework meets the standard requirements of efficiency and scalability. The accuracy of the framework can be improved by developing a precise and efficient knowledge base of the commonly used cybercrime terms. Moreover, the result can be improved by employing sophisticated techniques in the preprocessing step and by using a dataset that is predominantly malicious. The current version of WordNet contains a limited number of cybercrime related words and therefore needs to be extended to include more terms. Similarly, in order to extend the proposed framework to support languages other than English, the development of a WordNet-like lexical database, e.g., EuroWordNet, is necessary.

Chapter 11
Artificial Intelligence And Digital Forensics

Artificial intelligence (AI) is a well-established branch of computer science concerned with making machines smart enough to perform computationally large or complex tasks that normally require human intelligence; furthermore, it comprises a combination of technologies that can obtain insights and patterns from a massive amount of data which is a crucial element of forensic analysis. This chapter focuses on AI and its subfields: machine learning and deep learning—in general—and also details AI and data mining techniques pertaining to digital forensics. In highlighting the current shortcomings of prevailing approaches, we propose a new approach to offer a clearer insight into potential data, and/or detect variables of interest, as well as assess the future of digital forensics in the concluding section.

The concept of AI and its machine learning counterpart have been around since the 1950s, however serious studies into these topics have experienced a significant rise due to technological advances, and increase in the processing and computational power of smart digital devices, including computers and embedded tiny devices. Currently, AI is being applied to almost all areas of science; examples used in everyday life include autonomous vehicles, conversational bots, recommender systems, medical assistance, stock trading advisory capacities, etc. The applications of AI are increasing day by day, however, its potential to perform malicious activities are increasing at the same pace. This leads to the emergence of digital forensics for incident response and forensics investigation of cybercrimes.

Digital forensics comprises several steps to collect, preserve, and analyze pieces of evidence found in digital devices and storage media. Recently, smart devices have become popular due to their processing power and storage, with newly emerging devices that can store enormous amounts of personal and commercial data. However, the constant growth in the capacity of digital storage media has resulted in a significant volume of data to examine in forensic analyses, with current resources, both in terms of individuals and computers, being insufficient to meet demands. Therefore, with regards to forensic analysis, which requires intelligent

F. Iqbal et al., *Machine Learning for Authorship Attribution and Cyber
Forensics*, International Series on Computer Entertainment and Media
Technology, https://doi.org/10.1007/978-3-030-61675-5_11

analysis of large and complex data, AI forms a perfect option to combat the main analytical problems faced by digital forensics. AI can help automate some processes and more quickly flag content or present insights that would otherwise take investigators longer to uncover. During analysis, different groups in the data can be identified using pattern recognition. Pattern recognition relies on statistics or probabilistic reasoning (or both) and can be applied to images, text, or disk images to identify groups of interest during analysis and investigation. AI can identify such patterns in complex data in a much more accurate manner.

11.1 AI Techniques

There are many popular examples of data mining currently being utilized in the forensic domain. Data mining from a digital forensic perspective is twofold: data mining to reveal purposefully hidden information, and data mining to understand large complex datasets of information. Both of these can be applied to multiple digital forensic data subsets from a variety of devices (including computers, mobile phones, IoT devices) to extract information.

Knowledge discovery and data mining can be applied to extract useful information from a large collection of data such as past cases or police reports. Many decision support systems employ knowledge discovery and data mining techniques, e.g., COPLINK [52], FLINTS [164], CIDDS [165], etc. COPLINK is a very popular crime analytics platform that supports several tasks such as auditing crime data, sharing this historical data with external agencies, applying link analysis/attribute association, all of which can help identify the suspects and their suspicious activities. FLINT is a tool for police investigation and intelligence analysis. Analysts, investigators, and policymakers can use this tool to find the connections between people, crime, location, time, and evidence. Similarly, the Sri Lankan Police Service has developed the Crime Investigation Decision Support System: CIDSS. CIDSS an intelligent crime analysis system in which data mining tools and techniques are used for analysis of the historical and current crime data. The system is developed with geographical information that supports spatial analysis and provides an efficient solution to manual map generation methods. AI can also help to extract strong statistical evidence. As in the forensic analysis, narratives and arguments are supported with statistical pieces of evidence; using AI, graphical structures can be designed that support building scenarios and case studies and can be used to prove or disprove arguments. Mathematical and computational tools using AI can also help to build statistically relevant and significant pieces of evidence.

Machine learning (ML) is a specific subset of AI that trains a machine on how to learn and adapt. Using ML, a system can learn from experience and examples provided to reduce the time and resources required to accomplish a task. ML is capable of aiding digital investigators at different stages of the process. For its application in digital forensics, ML involves deductive and inductive learning [173]. Deductive learning uses ML to refine the knowledge source to keep it current and updated. By contrast, inductive learning applies ML to gather initial knowledge. In some techniques, both approaches are utilized to achieve results. With regards to ML, we can utilize three distinct machine learning models: Supervised, Unsupervised, and Reinforcement Learning. Supervised ML is the most common type of machine learning, whereby models are trained using labeled examples of past data to predict future events such as spam detection. In unsupervised learning, patterns are extracted from data with no pre-existing labels e.g. anomaly detection. In reinforcement learning, the system employs trial and error to come up with a solution to the problem.

Deep learning is a branch of ML which is based on artificial neural networks. Deep neural networks have multiple hidden layers, between the input and output layers. These models are trained using large sets of labeled data and network architectures that learn features directly from the data (without the need for manual feature extraction). Deep learning architectures have been successfully employed in the fields of computer vision, natural language processing (or NLP), drug discovery, recommendation systems, bioinformatics, etc. Recent achievements in deep learning have opened up many opportunities and have also led to widespread adoption and deployment of deep neural networks in security-critical systems. Due to the availability of cheap storage and cloud services, as well as the increasing adoption of IPv6, investigators are applying big data technology, cloud-based forensics services, as well as deep learning techniques to collect and efficiently analyze a large volume of heterogeneous evidence.

In the previous chapter, we discussed a framework for extracting criminal activities from the analysis of suspicious online messages. In this chapter, we will explore the applications of deep learning in forensic analysis and textual data processing. Due to availability, anonymity, and convenience, the Internet and a variety of digital devices have experienced a sharp rise in use by criminals to facilitate their offenses. Investigators must largely deal with textual data and large files such as e-mails, chat application logs, tweets, etc. The limited scale and size of e-mails and chat logs hamper statistical methods such as topic modeling, which are often unable to discover the themes of texts. Moreover, due to its unstructured nature and grammatical errors, traditional NLP techniques cannot be applied.

In forensic analysis, when there are large log files, the investigation starts by parsing the log files in which each field is identified and labeled based on the required information, e.g., timestamp, hostname, service name, IP address, and many other informational entries. Since the same concept is typically expressed by different terms and language styles, using keyword searches is not always effective. Previously, regular expressions and pyparsing have been utilized to process the large log files in an investigation. Regular expression is a sequence of characters that define a search pattern while pyparsing is a powerful alternative to regular expressions for parsing text into tokens. Pyparsing has more human-readable rules than regular expression; nonetheless, these approaches still require human input to define rules which is a tedious task and requires domain experts. Nerlogparser [92] an automatic tool for event log parsing for forensic investigators—has been proposed, which uses deep machine learning techniques, specifically the bi-directional long short-term memory networks, as the underlying architecture. It is a generic and fully automatic tool that does not require investigators to define any parsing rules and can parse various types of log files.

Unsupervised learning or clustering is a powerful AI technique where the investigators do not necessarily need to have initial knowledge or important leads beforehand to start an investigation. Clustering is defined as finding groups of objects in the data that share similarities, arranging and separating them from the other data objects. Text clustering also uses NLP to categorize unstructured text. Text clustering algorithms are divided into several types such as agglomerative clustering algorithms, partitioning algorithms, and standard parametric modeling-based methods. Agglomerative hierarchical clustering assigns each tweet as a singleton cluster, then groups pairs of clusters based upon similarities until a single cluster represents all groupings and tweets. It generates a cluster hierarchy in which the leaf nodes represent individual tweets, and the internal nodes correspond to the merged groups of clusters. The quality of the clustering algorithm is dependent on the features used in the clustering, so feature selection and representation are very critical processes. Vector space modeling represents documents as vectors in a common vector space, where each dimension of this vector represents a separate term. Word embeddings are used to represent tweets because they have shown a good generalization power for feature representation in many NLP tasks, e.g., named entity recognition, dependency parsing, text classification, etc. In NLP applications such as classification, clustering, etc.; similarity/distance measure is an important building block. There are several distance metrics for text vectors. Euclidean distance and Cosine distance are known to perform well in practice; however, they are unable to capture cluster similarities when the same concept is written using different words.

To summarize, deep learning techniques can be used to extract the required knowledge in order to learn and establish any possible relationships stored in huge data sets; furthermore, these data sets can be collected from historical crime data that authorities collect over time. In the domain of digital forensics, AI offers great opportunities for accurate data analysis and interpretation. However, current ML techniques encounter difficulty handling the myriad datasets and devices in modern society. The approach presented in the subsequent sections demonstrates the applicability of state-of-the-art deep learning techniques for detecting threat-intelligence from social media.

11.2   Deep Learning for Social Media Mining

In times of emergencies and events, people often use social media, such as Twitter, to break the news, exchange information with other users, mobilize and unite said users, and raise funds for victims. In such situations, efficient and effective actions are crucial for mitigating the disaster or emergency in question. Therefore, the early detection and collection of such information is imperative for taking quick and appropriate action. This information (which usually concerns the nature of incident, potential, and expected losses in terms of human lives and monetary losses, etc.), if successfully gathered, can help facilitate rescue operations, providing first aid and other such medical services, firefighting, etc. Furthermore, law enforcement agencies and emergency services, e.g., firefighters, paramedics, and rescue teams, could utilize this information to quickly assess the seriousness of an event and organize an appropriate response with increased efficiency and speed. Additionally, data gathered from social media platforms can help forensic experts in evidence collection and allow said investigators to contact the eyewitnesses of an event or incident. While the data obtained can serve several useful purposes, one of the major challenges lies in the processing of this large, recurrent, and noisy Twitter data to extract meaningful information. The objective of the proposed method in this chapter is to design and implement a novel automatic terror attack detection system that monitors the Twitter stream and, in case of a terror attack, extracts all related information shared on Twitter to alert law enforcement, emergency services, and the media. The proposed method employs word embedding to transform tweets into a vector space model, then utilizes Word Movers Distance (WMD) to cluster tweets for the detection of incidents. The architecture of the proposed system is shown in Fig. 11.1.

The following subsections present details of each component.

11.2.1 Tweet Crawler

Tweet crawler is used in this proposed system that "crawls" through the relevant tweets and stores them for future use. It uses the Twitter API and domain-specific keywords for the tweet search. Twitter offers APIs which have the ability to filter

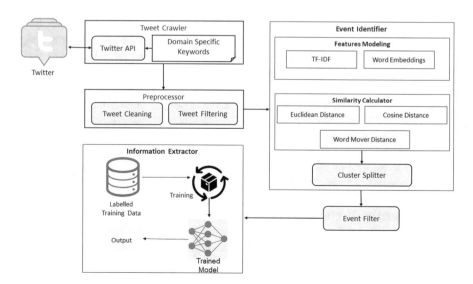

Fig. 11.1 Architecture of the proposed system

real time tweets using multiple filtering rules per connection. Tweets can be filtered out based upon a wide variety of attributes, e.g., keywords, geolocation, language, etc. The user can enter keywords related to terror attacks such as "blast", "explosion", "shooting", etc., to retrieve the latest relevant tweets. These can be expanded to include metadata useful for establishing timelines in an investigation. The filtering of informative tweets and classification of these tweets into meaningful categories or groups is critical for first responders and decision makers.

11.2.2   Preprocessing

Data collected from social media is often noisy and heterogeneous in nature. To prepare the Twitter stream for analysis, the preprocessing step removes all the mentions, URLs, special characters, and stop words from the tweets. For hashtags, it just removes "#" symbol from the start of the token. These can be individually adapted per investigation to improve accuracy, whereby the data is prepared and inspected prior to clustering. The tokenizer element handles the splitting of hashtags into valid words e.g. for the hashtag "#HighParkFire", it automatically converts into 'high', 'park', and 'fire'. After preprocessing, the collected tweets are passed to the tweet classifier. Additional functions of preprocessing include lemmatization, which is the reduction of words such as "am", "are", and "is" to a more common form such as "be", simplifying analysis of the data.

11.2.3   Event Identifier

Event identification is performed using the clustering method to group incoming tweets. In this approach, agglomerative hierarchical clustering has been chosen as it does not require the user to pre-specify the number of clusters, like in partition clustering. Tf-idf and Word Embedding are two of the most common methods in NLP to convert sentences to machine-readable code. Tf-idf has been explained in Chap. 3 in detail. Word embedding is a feature learning method that converts a word into an n-dimensional vector so that similar words like house and home will have the same numerical representation. It has the ability to capture the context of a word in a document. Several methods have been proposed to learn word embeddings from a large amount of unlabeled text. Word2Vec, proposed by Google [174] is the most popular method to learn word embeddings using shallow neural networks. It is a statistical method for efficiently learning a standalone word embedding from a body of text, by allowing the target word to range over the body of text and the context word to range over a context window. It is a combination of two models: Continuous bag of words (CBOW) and Skip-gram. Both are shallow neural networks that map words to the target variable. The CBOW model learns the embedding by predicting the current word based on its context, whereas the skip-gram model learns by predicting the words surrounding a given word. Global Vectors for Word Representation (GloVe) [175] is an unsupervised learning algorithm for vector representation that is able to preserve semantic and syntactic regularities in the text. It works like Word2Vec but is a count-based model that trains on the word co-occurrence counts and thus makes efficient use of statistics. fastText [176] is a library created by the Facebook Research Team for efficient learning of word representations. As compared to the Word2Vec model, it treats each word as being composed of character n-grams so that the vector for a word is made up of the sum of its character n-grams. It is able to generate better word embedding for rare and less commonly used vocabulary using its character n-gram.

To measure the similarity between two tweets, the method specifically proposes the use of the WMD [177]. WMD measures the similarity between two text documents in a meaningful way even if there is no commonality in the words of the two documents, leverages word embeddings, and defines the distances between two documents. As such, it can find a meaningful distance between the tweets written by different users in different linguistic styles. Moreover, it generates a good cluster representation in terms of tight and loose balance. After measuring the distance/similarity between tweets, the system merges the two most similar clusters at each iteration.

Hierarchical clustering does not require a number of clusters at the initialization phase. To obtain flat clusters, branches of the dendrogram can be cut at a specific level: a process referred to as tree cutting or dendrogram pruning. Cutting a hierarchy at a specific level gives a set of clusters while cutting at another level gives a different set of clusters; it depends on the application, data, and required granularity. A higher cutoff value results in looser clusters that can contain multiple events in the same cluster, while a lower cutoff value results in clean clusters but with a higher degree of event fragmentation. After clustering, all the identified clusters are passed to the event filter.

11.2.4 Event Filter

As Twitter data is noisy, it results in several small clusters with no useful information or outliers. To filter outliers, event filter counts the number of tweets in each cluster. If the total number of tweets in a cluster is less than a predefined threshold, then it is discarded.

11.2.5 Information Extractor

After identifying a terrorist attack, the proposed system processes all relevant tweets in the cluster to extract valuable information in order to analyze the situation and better plan for emergency response. Recurrent neural network (RNN) is an artificial neural network with loops that takes time and sequence into account and retains information of the previous state of the neural network. RNN can make use of context, recognize sequential patterns in the presence of sequential distortions, and can be used with different data types and representation. These unique properties of RNN make it an optimal choice for sequence labeling. Long short-term memory (LSTM) networks are a type of RNN architecture that can learn long-term dependencies. For a sequential labeling task, a LSTM model can take into account an infinite amount of context and eliminate the problem of limited context that applies to any feed-forward model. A bi-directional Long Short-Term Memory Recurrent Neural Network (bLSTM) is a combination of two LSTMs: one runs forward from "right to left", and one runs backward from "left to right".

To extract valuable information from a Twitter post, the problem of sequence labeling can be defined as follows:

For a given tweet t_i regarding a specific terrorist attack, the task is to assign a label to each word of the tweet such that for an observation sequence $w = (w^1, w^2, w^3, \cdots, w^n)$, the output is a sequence of labels $y = (y^1, y^2, y^3, \cdots, y^n)$. Figure 11.2 shows the bLSTM network designed to extract attack-related information. To combine the output of the forward and the backward layer, there are options of concatenation, summation, multiplication, and average. Each node in the input layer is connected with two separate hidden layers: one hidden layer processes the input sequence of features forward, and the other hidden later processes it backwards. The output layer has a size equal to the number of tags to identify. The SoftMax activation function has been used for the output layer.

After labeling tweets using the bLSTM model, the system processes the extracted information to generate reports for law enforcement and emergency services. These reports contain information such as the incident location, attack type and time, number of deaths, number of injuries, etc. As tweets are often randomly written by the general public, sometimes multiple entities are given under the same label. For example, there could be a different number of deaths or injuries as a result of the

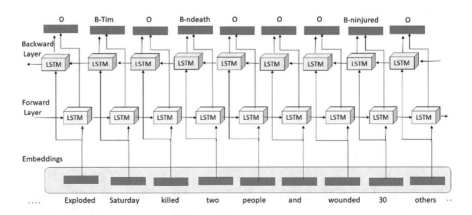

Fig. 11.2 bLSTM network for extraction of attack-related information

attack mentioned by Twitter users. For such cases, the system selects the range of those numbers from minimum to maximum in the report.

Transformers are introduced by Google in the paper "Attention Is All You Need" [178]. Transformers are sequence to sequence neural networks, designed to handle sequential data such as speech recognition, text-to-speech transformation, summarization, and translation with long term dependencies. Transformer models consist of encoder that takes an input sequence and maps it into a higher dimensional space (n-dimensional vector) and decoder that converts encoded sequence into an output sequence. In training the transformer model, the attention method comes into play at each step that decides which other parts of the sequence are important and assigns weights to them based on their importance. Transformer models use feed-forward layers and attention layers stacked on each other which enables them to handle input of variable size. Each layer can process in parallel. Transformers have shown remarkable performance on various NLP tasks such as machine translation [178], document generation [179], and syntactic parsing [180]. Transfer learning is another approach that uses knowledge learned from one dataset and applies it to various other problems. It has proven to be very effective in providing good accuracy with very small amounts of labeled data.

Bidirectional Encoder Representations from Transformers (BERT) [181] is a powerful model by Google AI trained on a large amount of unlabeled data. Unlike the other neural network structures for NLP, BERT is a bidirectional encoder representation from transformers that fully includes the bidirectional contextual features inside the model. Deep bidirectional representations are trained by incorporating both left and right contexts in all layers. These representations are learned by two unsupervised tasks on a large amount of unlabeled data. The first task is called masked LM, which masks some percentage of the input tokens randomly, and then predict those masked tokens. The second task is called the next sentence prediction, which is a binary prediction task. It is pre-trained for pair of two sentences—A and B—where 50% of the time B is the actual next sentence that follows A, and 50% of the time it is a random sentence from the body of text. It uses pre-training and fine-tuning to create state-of-the-art models for different NLP tasks such as question answering systems, sentiment analysis, and language inference. In the pre-training phase, the model is trained on a large amount of unlabeled text and for fine-tuning, the model is first initialized with the pre-trained parameters which are fine-tuned using a small amount of labeled data from the downstream task. Each downstream task has separate fine-tuned models.

Google has released two pre-trained BERT models: BERT-Base and BERT-Large. BERT-Base is trained using 12 transformer layers, 768 hidden nodes, 12 attention heads, and 110M parameters whereas BERT Large has 24 transformer layers, 1024 hidden nodes, 16 attention heads, and 340M parameters. Recent language models are very powerful and are able to perform in many different fields. Some recent work leveraged language models in the forensic domain. Chiyu et al. [182] used the BERT model to detect age, language variety, and gender based on posted language in the context of the Arabic author profiling and deception detection. They fine-tuned BERT using labeled data for downstream tasks. Moreover,

Georgios et al. [183] examined the use of pre-trained language models for cross-domain authorship attribution. Their results demonstrated that pre-trained language models like BERT perform well and their performance remains stable in different scenarios.

In this approach, classification is performed by fine-tuning the BERT model on domain-specific data. Two classifiers are trained for tweets collected during a crisis. The first classifier classifies the tweets to separate informative tweets from non-informative ones. Classification of tweets removes noise and non-informative posts and tweets become cleaner for assessing the situation in terms of specific categories established for the event. The second classifier performs fine-grained classification of informative tweets into six categories based on the type of information they contained. In the case of an attack, these categories would be "Affected individuals", "Caution and advice", "Donations and volunteering", "Infrastructure and utilities", "Sympathy and support", and "Other useful information".

As a result, the system information related to the crisis can be easily classified using a small amount of labeled data, which consequently will help address the varying needs of different individuals and/or agencies involved in a crisis. This chapter has demonstrated one example of social media data mining and the benefits deep learning can bring to the forensic field in terms of data cleaning, filtering, and feature extraction. The relevance of the data gathered depends on the underlying algorithms, neural network set up and historical data, but the prevailing capabilities of analytics can offer us insight and provide structure to this mass collection of data. While the examples within this chapter convey pattern recognition of various bodies of text, pattern recognition becomes more challenging and complex when presented with other data types. The use of AI in digital forensics is still at a very early stage, but it does have a lot to offer the digital forensics community. Social media data and future data collection can involve many disparate and variable types of data, so future algorithms should also take these variables into account. With that in mind, the use of more complex pattern recognition and data mining techniques will have a great impact on future studies into this topic.

11.3 Future Applications and Implications for Digital Forensics

One of the major challenges facing digital forensics is the creation of tools and techniques to analyze the increasingly diverse devices and bulk of data generated. AI in digital forensics has a key role in extending and advancing the capabilities of currently used tools, yet there is a need to prepare for approaches that can handle even larger variations of data sizes and attributes. As previously stated, every individual with an online presence has a large digital trail of information that can be analyzed and interpreted; many of which can span a significant period of time. The scale of this data is simply too much for human analysis alone; as such, AI is required for collection and analyses of logs and evidence. The implementation of AI holds the potential to dramatically change the field of digital forensics.

The use of behavioral and data sciences to understand how people interact with devices and technologies will be key in establishing authorship and intent in criminal situations. We have previously demonstrated the use of AI applications to combating crime online via the application of machine learning, text mining, and text classification from suspicious online messages. Could this collection of data, or historical 'big data' be used for preventative measures, i.e., could it notice behavior indicative of someone who may be considering committing an offense? These compendia of chapters have explored using AI for authorship identification and attribution, and for identifying other players in online crimes, but it would be worth considering using this collection of data for creating profiles of online offenders and analyzing behavior potentially indicative of future offenses. Current approaches of AI in forensics involve live scraping or post incident analysis; can this data be used for forecasting models and feedback into the systems, like the underlying neural networks?

Additionally, the handling of forensic data receives a lot of attention—as such, it can be construed that Blockchain-based solutions for maintaining and tracing digital forensics chain of custody will have potential ramifications; e.g. taking hashes of digital evidence and recording them securely on blockchain through smart contracts. Details such as the location, time, and date of crime scenes can also be recorded on the blockchain as a way to maintain the integrity of evidence and present a clear chain of command in a collaborative investigation. Additionally, Blockchain can make AI more coherent and understandable. Decisions taken by AI systems can be difficult for humans to comprehend and even harder to explain in a court of law. Evidence collected using AI should be explainable to a judge and court of law, and analytic processes should allow for rollback, or require full documentation detailing step by step processes, including all technical aspects of the underlying algorithms and models [184]. Blockchain can support and facilitate these processes by helping to track the problem-solving process and understand decisions. With the ability to trace and determine why decisions are made, recording all data and variables that are used in decision making under machine learning will help AI eventually become more explainable and auditable within the legal process.

Modern digital societies are frequently subject to cybercriminal activities and fraud, increasing economic losses for both individuals and organisations. Consequently, future forensics tools should be designed and developed to support heterogeneous investigations, preserve privacy, and offer scalability [185]. It is important to note that the role of human intervention is an important player in the utilization of these new technologies in the forensic process. The underlying algorithms and models require training of data and reference models; the idea being "the more data the smarter the model". For accuracy, human intervention should be used to check and maintain the integrity of the AI models, ensuring that the model and parameters fit a specific use case. Reflecting on the applications conveyed, it is hard to create an approach for all types of crimes, as cybercrimes cover a wide variety of offenses. Data sample factors should also be taken into consideration, as dataset size poses challenges to replicability, and sampling factors can have different meanings per scenario.

In order to deal with this fundamental change in evidence, it is essential that the techniques used in these investigations must change and develop in order to deal with these effectively. As noted previously, investigators face an increasingly large and complicated pool of data to sift through, with less time and finances to cover these increased demands. For the advancement in this paradigm, there is a need for a change of forensic readiness in order to combat future criminal activities and assist forensic analyses.

References

1. B. Nelson, A. Phillips, C. Steuart, *Guide to Computer Forensics and Investigations* (Cengage Learning, 2014)
2. M.E. Whitman, H.J. Mattord, *Principles of Information Security* (Cengage Learning, 2011)
3. M. Ciampa, *Security+ Guide to Network Security Fundamentals* (Cengage Learning, 2012)
4. T. Olovsson, A structured approach to computer security (1992)
5. World Internet Users Statistics and 2018 World Population Stats (2018). [Online]. https://www.internetworldstats.com/stats.htm
6. A. Smith, Cell Internet Use 2012 (2012). [Online]. https://www.pewresearch.org/internet/2012/06/26/cell-internet-use-2012/. Accessed 3 May 2020
7. The Mobile Economy 2018. [Online]. https://www.gsma.com/mobileeconomy/wp-content/uploads/2018/05/The-Mobile-Economy-2018.pdf
8. Ponemon Institute LLC, Cost of Data Breach Study | IBM Security (2018). [Online]. https://www.ibm.com/security/data-breach
9. Accenture, 2017 COST OF CYBER CRIME STUDY (2017). [Online]. https://www.accenture.com/us-en/insight-cost-of-cybercrime-2017
10. Ponemon Institute LLC Research Report, 2015 Cost of Data Breach Study: Global Analysis (2015). [Online]. Available: https://nhlearningsolutions.com/Portals/0/Documents/2015-Cost-of-Data-Breach-Study.PDF. Accessed 12 Nov 2020
11. J.P. Lawler, H. Howell-Barber, A framework model for a software-as-a-service (SaaS) strategy, in *Encyclopedia of Information Science and Technology*, 3rd edn. (IGI Global, 2015), pp. 1024–1032
12. The Global State of Information Security Survey 2018 (2018). [Online]. https://www.pwc.com/us/en/services/consulting/cybersecurity/library/information-security-survey.html
13. B. Wire, Dell Global Security Survey: Organizations Overlook Powerful New Unknown Threats, Despite Significant Costs. [Online]. https://www.businesswire.com/news/home/20140220005281/en/Dell-Global-Security-Survey-Organizations-Overlook-Powerful
14. T. Brewster, Certificate authority confirms hack after Gmail attack. [Online]. https://www.itpro.co.uk/635833/certificate-authority-confirms-hack-after-gmail-attack
15. J. Fruhlinger, What is WannaCry ransomware, how does it infect, and who was responsible (2017)
16. S. Ragan, Code Spaces forced to close its doors after security incident, *CSO June*, vol. 18 (2014)
17. Pro OnCall Technologies Resources Library—IT Whitepapers & eBooks (2018). [Online]. https://prooncall.com/resources/

18. Information is beautiful "World's Biggest Data Breaches & Hacks" [Online]. https://www.informationisbeautiful.net/visualizations/worlds-biggest-data-breaches-hacks/. Accessed 12 Nov 2020

19. HP News—10-08-HP Reveals Cost of Cybercrime Escalates 78 Percent, Time to Resolve Attacks More Than Doubles (2017). [Online]. http://www8.hp.com/us/en/hp-news/press-release.html?id=1501128#.W0gwF9JKiM8

20. G. Tsakalidis, K. Vergidis, A systematic approach toward description and classification of cybercrime incidents. IEEE Trans. Syst. Man Cybern. Syst. **99**, 1–20 (2017)

21. 2014 Internet Crime Report. [Online]. https://pdf.ic3.gov/2014_IC3Report.pdf

22. K.M. Finklea, C.A. Theohary, Cybercrime: conceptual issues for congress and US law enforcement (2012)

23. Convention on Cybercrime. [Online]. https://www.coe.int/en/web/conventions/full-list/-/conventions/treaty/185. Accessed 5 May 2020

24. D.L. Shinder, M. Cross, *Scene of the Cybercrime* (Elsevier, 2008)

25. 2015 Internet Crime Report. [Online]. https://pdf.ic3.gov/2015_IC3Report.pdf

26. Mississippi woman charged in identity theft over 2 years. [Online]. https://apnews.com/6927169a2b53d4e47c0227d5246eb258. Accessed 12 May 2020

27. Justice Department warns of disaster fraud after Irma, Harvey (2017). [Online]. https://www.reuters.com/article/us-storm-irma-fraud/justice-department-warns-of-disaster-fraud-after-irma-harvey-idUSKCN1BP2MX

28. Cisco, What Is Phishing? [Online]. https://www.cisco.com/c/en/us/products/security/email-security/what-is-phishing.html#~how-phishing-works

29. PwC, The Global State of Information Security® Survey 2015 (2015). [Online]. https://www.pwc.ru/en/publications/information-security-survey1.html

30. R. Anderson et al., Measuring the cost of cybercrime, in *The Economics of Information Security and Privacy* (Springer, 2013), pp. 265–300

31. F. Howard, Exploring the Blackhole exploit kit, Sophos Tech. Pap. (2012)

32. IntSights Cyber Intelligence, IntSights Report Shows Banking and Financial Services Sector Continues to Be Targeted by External Threat Actors (2019). [Online]. https://www.prnewswire.com/il/news-releases/intsights-report-shows-banking-and-financial-services-sector-continues-to-be-targeted-by-external-threat-actors-300839696.html

33. Combating Computer Crime (2017). [Online]. https://www.hg.org/legal-articles/combating-computer-crime-31034

34. RIG (2016). [Online]. https://www.cyber.nj.gov/threat-profiles/exploit-kit-variants/rig. Accessed 13 Oct 2018

35. P. Richards, Nation state attacks—the cyber cold war gets down to business (2018). [Online]. https://www.csoonline.com/article/3268976/cyberwarfare/nation-state-attacks-the-cyber-cold-war-gets-down-to-business.html

36. D. Coats, Worldwide Threat Assessment of US Intelligence Community (2018). [Online]. https://www.dni.gov/files/documents/Newsroom/Testimonies/2018-ATA%2D%2D-Unclassified-SSCI.pdf

37. M. Shirer, Worldwide Public Cloud Services Revenue Growth Remains Strong Through the First Half of 2017, According to IDC (2017). [Online]. https://www.businesswire.com/news/home/20171106005140/en/Worldwide-Public-Cloud-Services-Revenue-Growth-Remains

38. New Report on the State of Phishing Attacks from Wombat Security Shows Significant Increases Year over Year (2016)

39. (APWG), Phishing Activity Trends Report: Unifying the Global Response To Cybercrime (2017). [Online]. https://docs.apwg.org/reports/apwg_trends_report_q4_2016.pdf

40. A. Abbasi, H. Chen, J.F. Nunamaker, Stylometric identification in electronic markets: scalability and robustness. J. Manag. Inf. Syst. **25**(1), 49–78 (2008)

41. H. Chen et al., Crime data mining: an overview and case studies, in *Proceedings of the 2003 Annual National Conference on Digital Government Research* (2003), pp. 1–5

42. R.C. der Hulst, Introduction to Social Network Analysis (SNA) as an investigative tool. Trends Organ Crime **12**(2), 101–121 (2009)

43. First Amendment And The Media 'Encyclopedia of Communication and Information' (2002). [Online]. https://www.encyclopedia.com/media/encyclopedias-almanacs-transcripts-and-maps/first-amendment-and-media

44. K.C. Darrell Etherington, Large DDoS attacks cause outages at Twitter, Spotify, and other sites (2016). [Online]. https://techcrunch.com/2016/10/21/many-sites-including-twitter-and-spotify-suffering-outage/

45. Oregon Woman Loses $400,000 to Nigerian E-Mail Scam (2008). [Online]. http://www.foxnews.com/story/2008/11/17/oregon-woman-loses-400000-to-nigerian-e-mail-scam.html

46. Scam Victim Stories, Scammer's Exposed (2017). [Online]. https://scammer419.wordpress.com/scam-victim-stories/

47. N. Chou, R. Ledesma, Y. Teraguchi, J.C. Mitchell et al., Client-side defense against web-based identity theft, in *NDSS* (2004)

48. C.E.H. Chua, J. Wareham, Fighting internet auction fraud: an assessment and proposal. Computer (Long. Beach. Calif) **37**(10), 31–37 (2004)

49. G.-F. Teng, M.-S. Lai, J.-B. Ma, Y. Li, E-mail authorship mining based on SVM for computer forensic, in *Proceedings of 2004 International Conference on Machine Learning and Cybernetics*, vol. 2 (2004), pp. 1204–1207

50. Forensic ToolKit. [Online]. https://accessdata.com/products-services/forensic-toolkit-ftk. Accessed 5 May 2020

51. Encase. [Online]. http://www.guidancesoftware.com/. Accessed 5 May 2020

52. Data Warehousing—Coplink*/BorderSafe/RISC. [Online]. https://eller.arizona.edu/departments-research/centers-labs/artificial-intelligence/research/previous/coplink. Accessed 5 May 2020

53. Paraben's E3: EMX. [Online]. https://www.paraben.com/products/e3-emx. Accessed 5 May 2020

54. S.J. Stolfo, S. Hershkop, Email mining toolkit supporting law enforcement forensic analyses, in *Proceedings of the 2005 National Conference on Digital Government Research* (2005), pp. 221–222

55. S. Argamon, M. Šarić, S.S. Stein, Style mining of electronic messages for multiple authorship discrimination: first results, in *Proceedings of the Ninth ACM SIGKDD International Conference on Knowledge Discovery and Data Mining* (2003), pp. 475–480

56. M. Koppel, J. Schler, S. Argamon, Computational methods in authorship attribution. J. Am. Soc. Inf. Sci. Technol. **60**(1), 9–26 (2009)

57. H. Baayen, H. Van Halteren, F. Tweedie, Outside the cave of shadows: using syntactic annotation to enhance authorship attribution. Liter. Linguist. Comput. **11**(3), 121–132 (1996)

58. J.F. Burrows, Word-patterns and story-shapes: the statistical analysis of narrative style. Liter. Linguist. Comput. **2**(2), 61–70 (1987)

59. F. Mosteller, D.L. Wallace, *Applied Bayesian and Classical Inference: The Case of the Federalist Papers* (Springer Science & Business Media, 2012)

60. J.F. Burrows, 'An ocean where each kind...': statistical analysis and some major determinants of literary style. Comput. Hum. **23**(4–5), 309–321 (1989)

61. R.S. Forsyth, D.I. Holmes, Feature-finding for test classification. Liter. Linguist. Comput. **11**(4), 163–174 (1996)

62. O. De Vel, Mining e-mail authorship, in *Proc. Workshop on Text Mining, ACM International Conference on Knowledge Discovery and Data Mining (KDD'2000)* (2000)

63. R. Zheng, J. Li, H. Chen, Z. Huang, A framework for authorship identification of online messages: writing-style features and classification techniques. J. Am. Soc. Inf. Sci. Technol. **57**(3), 378–393 (2006)

64. F. Iqbal, R. Hadjidj, B.C.M. Fung, M. Debbabi, A novel approach of mining write-prints for authorship attribution in e-mail forensics. Digit. Investig. **5**, S42–S51 (2008)

65. O. De Vel, A. Anderson, M. Corney, G. Mohay, Mining e-mail content for author identification forensics. ACM SIGMOD Rec. **30**(4), 55–64 (2001)

66. J. Foertsch, The impact of electronic networks on scholarly communication: avenues for research. Discourse Process. **19**(2), 301–328 (1995)
67. V. Benjamin, W. Chung, A. Abbasi, J. Chuang, C.A. Larson, H. Chen, Evaluating text visualization for authorship analysis. Secur. Inform. **3**(1), 10 (2014)
68. T.C. Mendenhall, The characteristic curves of composition. Science **9**(214), 237–249 (1887)
69. J. Schroeder, J. Xu, H. Chen, M. Chau, Automated criminal link analysis based on domain knowledge. J. Am. Soc. Inf. Sci. Technol. **58**(6), 842–855 (2007)
70. J. Allan, J. Carbonell, G. Doddington, J. Yamron, Y. Yang et al., Topic detection and tracking pilot study: final report, in *Proceedings of the DARPA Broadcast News Transcription and Understanding Workshop*, vol. 1998 (1998), pp. 194–218
71. R. Barzilay, N. Elhadad, Inferring strategies for sentence ordering in multidocument news summarization. J. Artif. Intell. Res. **17**, 35–55 (2002)
72. R. Barzilay, K.R. McKeown, Sentence fusion for multidocument news summarization. Comput. Linguist. **31**(3), 297–328 (2005)
73. D. Das, A.F.T. Martins, A survey on automatic text summarization. Liter. Surv. Lang. Stat. II Course C. **4**, 192–195 (2007)
74. M. White, T. Korelsky, C. Cardie, V. Ng, D. Pierce, K. Wagstaff, Multidocument summarization via information extraction, in *Proceedings of the First International Conference on Human Language Technology Research* (2001), pp. 1–7
75. N. Chinchor, Overview of MUC-7, in *Seventh Message Understanding Conference (MUC-7): Proceedings of a Conference Held in Fairfax, Virginia, April 29–May 1, 1998* (1998)
76. E. Minkov, R.C. Wang, W.W. Cohen, Extracting personal names from email: applying named entity recognition to informal text, in *Proceedings of the Conference on Human Language Technology and Empirical Methods in Natural Language Processing* (2005), pp. 443–450
77. G. Wang, H. Chen, H. Atabakhsh, Automatically detecting deceptive criminal identities. Commun. ACM **47**(3), 70–76 (2004)
78. V.R. Carvalho, W.W. Cohen, Learning to extract signature and reply lines from email, in *Proceedings of the Conference on Email and Anti-Spam*, vol. 2004 (2004)
79. H. Chen, W. Chung, J. Qin, E. Reid, M. Sageman, G. Weimann, Uncovering the dark Web: a case study of Jihad on the Web. J. Am. Soc. Inf. Sci. Technol. **59**(8), 1347–1359 (2008)
80. A. Pons-Porrata, R. Berlanga-Llavori, J. Ruiz-Shulcloper, Topic discovery based on text mining techniques. Inf. Process. Manag. **43**(3), 752–768 (2007)
81. F. Sebastiani, Machine learning in automated text categorization. ACM Comput. Surv. **34**(1), 1–47 (2002)
82. N. Pendar, Toward spotting the pedophile telling victim from predator in text chats, in *International Conference on Semantic Computing, 2007. ICSC 2007* (2007), pp. 235–241
83. E. Elnahrawy, Log-based chat room monitoring using text categorization: a comparative study, in *The International Conference on Information and Knowledge Sharing, US Virgin Islands* (2002)
84. H. Dong, S. Cheung Hui, Y. He, Structural analysis of chat messages for topic detection. Online Inf. Rev. **30**(5), 496–516 (2006)
85. T. Kolenda, L.K. Hansen, J. Larsen, Signal detection using ICA: application to chat room topic spotting, in *Third Int. Conf. Indep. Compon. Anal. Blind Source Sep.* (2001), no. 1, pp. 540–545
86. Ö. Özyurt, C. Köse, Chat mining: automatically determination of chat conversations' topic in Turkish text based chat mediums. Expert Syst. Appl. **37**(12), 8705–8710 (2010)
87. Y. Zhang, N. Zincir-Heywood, E. Milios, Narrative text classification for automatic key phrase extraction in web document corpora, in *Proceedings of the 7th Annual ACM International Workshop on Web Information and Data Management* (2005), pp. 51–58
88. R. Xiong, J. Donath, PeopleGarden: creating data portraits for users, in *Proceedings of the 12th Annual ACM Symposium on User Interface Software and Technology* (1999), pp. 37–44
89. J. Bengel, S. Gauch, E. Mittur, R. Vijayaraghavan, Chattrack: chat room topic detection using classification, in *International Conference on Intelligence and Security Informatics* (2004), pp. 266–277

90. G. Salton, M.J. McGill, Introduction to modern information retrieval (1986)
91. H. Chen, W. Chung, J.J. Xu, G. Wang, Y. Qin, M. Chau, Crime data mining: a general framework and some examples. Computer (Long. Beach. Calif). **37**(4), 50–56 (2004)
92. Y. Xiang, M. Chau, H. Atabakhsh, H. Chen, Visualizing criminal relationships: comparison of a hyperbolic tree and a hierarchical list. Decis. Support. Syst. **41**(1), 69–83 (2005)
93. E. Frank, M.A. Hall, I.H. Witten, *The WEKA Workbench. Online Appendix for "Data Mining: Practical Machine Learning Tools and Techniques"* (Morgan Kaufmann, 2016)
94. I.H. Witten, E. Frank, M.A. Hall, C.J. Pal, *Data Mining: Practical Machine Learning Tools and Techniques* (Morgan Kaufmann, 2016)
95. E. Frank, S. Kramer, Ensembles of nested dichotomies for multi-class problems, in *Proceedings of the Twenty-First International Conference on Machine Learning* (2004), p. 39
96. J.R. Quinlan, Induction of decision trees. Mach. Learn. **1**(1), 81–106 (1986)
97. M.D. Buhmann, *Radial Basis Functions: Theory and Implementations*, vol 12 (Cambridge University Press, 2003)
98. S.E. Robertson, K.S. Jones, Relevance weighting of search terms. J. Am. Soc. Inf. Sci. **27**(3), 129–146 (1976)
99. J. Pearl, Bayesian networks: a model of self-activated memory for evidential reasoning, in *Proceedings of the 7th Conference of the Cognitive Science Society, 1985* (1985), pp. 329–334
100. Sample Weka dataset. [Online]. https://storm.cis.fordham.edu/~gweiss/data-mining/datasets. html. Accessed 5 May 2020
101. H.C. Lee, T. Palmbach, M.T. Miller, *Henry Lee's Crime Scene Handbook* (Academic, 2001)
102. H. Jones, J.H. Soltren, Facebook: threats to privacy. Proj. MAC MIT Proj. Math. Comput. **1**, 1–76 (2005)
103. C. Eoghan, Digital evidence and computer crime, in *Forensic Sci. Comput. Internet. Op. Cit* (2004)
104. K.-K.R. Choo, R.G. Smith, R. McCusker, K.-K.R. Choo, *Future Directions in Technology-Enabled Crime: 2007-09* (Citeseer, 2007)
105. S.Ó. Ciardhuáin, An extended model of cybercrime investigations. Int. J. Digit. Evid. **3**(1), 1–22 (2004)
106. M. Bhattacharyya, S. Hershkop, E. Eskin, Met: an experimental system for malicious email tracking, in *Proceedings of the 2002 Workshop on New Security Paradigms* (2002), pp. 3–10
107. Discovering Email Header Forensic Analysis! (2017). [Online]. http://www.xploreforensics. com/blog/email-header-forensic-analysis.html. Accessed 5 May 2020
108. D.P. Chris et al., Another stemmer. ACM SIGIR Forum **24**(3), 56–61 (1990)
109. M.F. Porter, An algorithm for suffix stripping. Program **14**(3), 130–137 (1980)
110. A. Abbasi, H. Chen, Writeprints: a stylometric approach to identity-level identification and similarity detection in cyberspace. ACM Trans. Inf. Syst. **26**(2), 7 (2008)
111. T. Joachims, Text categorization with support vector machines: learning with many relevant features, in *European Conference on Machine Learning* (1998), pp. 137–142
112. G. Salton, Automatic Text Processing: The Transformation, Analysis, and Retrieval Of (Read. Addison-Wesley, 1989)
113. R.P. Rippmann, An introduction to computing with Neural Networks. IEEE ASSP Mag. **4**(2), 4–22 (1987)
114. R. Agrawal, J. Gehrke, D. Gunopulos, P. Raghavan, *Automatic Subspace Clustering of High Dimensional Data for Data Mining Applications*, vol. 27, no. 2 (ACM, 1998)
115. H. Li, D. Shen, B. Zhang, Z. Chen, Q. Yang, Adding semantics to email clustering, in *Sixth International Conference on Data Mining, 2006. ICDM'06* (2006), pp. 938–942
116. R. Zheng, Y. Qin, Z. Huang, H. Chen, Authorship analysis in cybercrime investigation, in *International Conference on Intelligence and Security Informatics* (2003), pp. 59–73
117. J. Rudman, The state of authorship attribution studies: some problems and solutions. Comput. Hum. **31**(4), 351–365 (1997)
118. S. Argamon, M. Koppel, G. Avneri, Routing documents according to style, in *First International Workshop on Innovative Information Systems* (1998), pp. 85–92

119. D.I. Holmes, The evolution of stylometry in humanities scholarship. Liter. Linguist. Comput. **13**(3), 111–117 (1998)

120. G.U. Yule, On sentence-length as a statistical characteristic of style in prose: with application to two cases of disputed authorship. Biometrika **30**(3/4), 363–390 (1939)

121. W.W. Greg, The statistical study of literary vocabulary. JSTOR 291–293 (1944)

122. F. Mosteller, D. Wallace, Inference and disputed authorship: the federalist (1964)

123. E. Stamatatos, N. Fakotakis, G. Kokkinakis, Automatic text categorization in terms of genre and author. Comput. Linguist. **26**(4), 471–495 (2000)

124. M. Corney, O. De Vel, A. Anderson, G. Mohay, Gender-preferential text mining of e-mail discourse, in *18th Annual Proceedings Computer Security Applications Conference, 2002* (2002), pp. 282–289

125. M. Gamon, Linguistic correlates of style: authorship classification with deep linguistic analysis features, in *Proceedings of the 20th International Conference on Computational Linguistics* (2004), p. 611

126. J. Diederich, J. Kindermann, E. Leopold, G. Paass, Authorship attribution with support vector machines. Appl. Intell. **19**(1–2), 109–123 (2003)

127. L.M. Manevitz, M. Yousef, One-class SVMs for document classification. J. Mach. Learn. Res. **2**, 139–154 (2001)

128. H. Van Halteren, Author verification by linguistic profiling: an exploration of the parameter space. ACM Trans. Speech Lang. Process. **4**(1), 1 (2007)

129. T. Kucukyilmaz, B.B. Cambazoglu, C. Aykanat, F. Can, Chat mining: predicting user and message attributes in computer-mediated communication. Inf. Process. Manag. **44**(4), 1448–1466 (2008)

130. Y. Zhao, J. Zobel, Effective and scalable authorship attribution using function words, in *Asia Information Retrieval Symposium* (2005), pp. 174–189

131. O. De Vel, A.M. Anderson, M.W. Corney, G.M. Mohay, Multi-topic e-mail authorship attribution forensics (2001)

132. M. Koppel, S. Argamon, A.R. Shimoni, Automatically categorizing written texts by author gender. Liter. Linguist. Comput. **17**(4), 401–412 (2002)

133. O.Y. de Vel, M.W. Corney, A.M. Anderson, G.M. Mohay, Language and gender author cohort analysis of e-mail for computer forensics (2002)

134. J. Novak, P. Raghavan, A. Tomkins, Anti-aliasing on the web, in *Proceedings of the 13th International Conference on World Wide Web* (2004), pp. 30–39

135. J.R. Quinlan et al., Learning with continuous classes, in *5th Australian Joint Conference on Artificial Intelligence*, vol. 92 (1992), pp. 343–348

136. S.L. Salzberg, C4. 5: programs for machine learning by j. ross quinlan. morgan kaufmann publishers, inc., 1993. Mach. Learn. **16**(3), 235–240 (1994)

137. N. Cristianini, J. Shawe-Taylor, *An Introduction to Support Vector Machines* (Cambridge University Press, Cambridge, 2000)

138. M. Van Uden, Rocchio: relevance feedback in learning classification algorithms, in *Proceedings of the ACM SIGIR Conference* (1998)

139. R. Agrawal, T. Imieliński, A. Swami, Mining association rules between sets of items in large databases. ACM SIGMOD Rec **22**(2), 207–216 (1993)

140. F. Iqbal, H. Binsalleeh, B.C.M. Fung, M. Debbabi, A unified data mining solution for authorship analysis in anonymous textual communications. Inf. Sci. (NY) **231**, 98–112 (2013)

141. F.J. Tweedie, R.H. Baayen, How variable may a constant be? Measures of lexical richness in perspective. Comput. Hum. **32**(5), 323–352 (1998)

142. J. Han, J. Pei, Mining frequent patterns by pattern-growth: methodology and implications. ACM SIGKDD Explor. Newsl. **2**(2), 14–20 (2000)

143. M.J. Zaki, Scalable algorithms for association mining. IEEE Trans. Knowl. Data Eng. **12**(3), 372–390 (2000)

144. B.C.M. Fung, K. Wang, M. Ester, Hierarchical document clustering using frequent itemsets, in *Proceedings of the 2003 SIAM International Conference on Data Mining* (2003), pp. 59–70
145. J.D. Holt, S.M. Chung, Efficient mining of association rules in text databases, in *Proceedings of the Eighth International Conference on Information and Knowledge Management* (1999), pp. 234–242
146. S.J. Stolfo, G. Creamer, S. Hershkop, A temporal based forensic analysis of electronic communication, in *Proceedings of the 2006 International Conference on Digital Government Research* (2006), pp. 23–24
147. A. Kulkarni, T. Pedersen, Name discrimination and email clustering using unsupervised clustering and labeling of similar contexts, in *IICAI* (2005), pp. 703–722
148. Enron Email Dataset. [Online]. https://www.cs.cmu.edu/~enron/. Accessed 5 May 2020
149. F. Iqbal, H. Binsalleeh, B.C.M. Fung, M. Debbabi, Mining writeprints from anonymous e-mails for forensic investigation. Digit. Investig. **7**(1–2), 56–64 (2010)
150. J.A. Hartigan, M.A. Wong, Algorithm AS 136: a k-means clustering algorithm. J. R. Stat. Soc. Ser. C (Appl. Stat.) **28**(1), 100–108 (1979)
151. A.P. Dempster, N.M. Laird, D.B. Rubin, Maximum likelihood from incomplete data via the EM algorithm. J. R. Stat. Soc. Ser. B **39**(1), 1–22 (1977)
152. B. Larsen, C. Aone, Fast and effective text mining using linear-time document clustering, in *Proceedings of the Fifth ACM SIGKDD International Conference on Knowledge Discovery and Data Mining* (1999), pp. 16–22
153. R. Thomson, T. Murachver, Predicting gender from electronic discourse. Br. J. Soc. Psychol. **40**(2), 193–208 (2001)
154. A. Martin, M. Przybocki, The NIST speaker recognition evaluation series, in *Natl. Inst. Stand. Technol. Web site* (2009)
155. Y. Freund, R.E. Schapire et al., Experiments with a new boosting algorithm, in *Icml*, vol. 96 (1996), pp. 148–156
156. J. Su, H. Zhang, C.X. Ling, S. Matwin, Discriminative parameter learning for bayesian networks, in *Proceedings of the 25th International Conference on Machine Learning* (2008), pp. 1016–1023
157. N. Friedman, D. Geiger, M. Goldszmidt, Bayesian network classifiers. Mach. Learn. **29**(2–3), 131–163 (1997)
158. J.C. Platt, 12 fast training of support vector machines using sequential minimal optimization, Adv. Kernel Methods 185–208 (1999)
159. C.J.C. Burges, A tutorial on support vector machines for pattern recognition. Data Min. Knowl. Discov. **2**(2), 121–167 (1998)
160. B.L.W.H.Y. Ma, B. Liu, Integrating classification and association rule mining, in *Proceedings of the Fourth International Conference on Knowledge Discovery and Data Mining* (1998)
161. X. Yin, J. Han, CPAR: classification based on predictive association rules, in *Proceedings of the 2003 SIAM International Conference on Data Mining* (2003), pp. 331–335
162. W. Li, J. Han, J. Pei, CMAR, Accurate and efficient classification based on multiple classassociation rules (2001)
163. F. Thabtah, P. Cowling, Y. Peng, MCAR: multi-class classification based on association rule, in *The 3rd ACS/IEEE International Conference on Computer Systems and Applications, 2005* (2005), p. 33
164. F. Coenen, G. Goulbourne, P. Leng, Tree structures for mining association rules. Data Min. Knowl. Discov. **8**(1), 25–51 (2004)
165. J. Han, J. Pei, Y. Yin, Mining frequent patterns without candidate generation. ACM SIGMOD Rec **29**(2), 1–12 (2000)
166. M.R. Schmid, F. Iqbal, B.C.M. Fung, E-mail authorship attribution using customized associative classification. Digit. Investig. **14**, S116–S126 (2015)
167. W. Li, *Classification Based on Multiple Association Rules* (Simon Fraser University, 2001)

168. E. Alfonseca, S. Manandhar, An unsupervised method for general named entity recognition and automated concept discovery, in *Proceedings of the 1st International Conference on General WordNet, Mysore, India* (2002), pp. 34–43
169. V.A. Yatsko, T.N. Vishnyakov, A method for evaluating modern systems of automatic text summarization. Autom. Doc. Math. Linguist. **41**(3), 93–103 (2007)
170. M.-H. Antoni-Lay, G. Francopoulo, L. Zaysser, A generic model for reuseable lexicons: the Genelex project (1994)
171. S.D. Kamvar, D. Klein, C.D. Manning, Interpreting and extending classical agglomerative clustering algorithms using a model-based approach (2002)
172. J. Heer, S.K. Card, J.A. Landay, Prefuse: a toolkit for interactive information visualization, in *Proceedings of the SIGCHI Conference on Human Factors in Computing Systems* (2005), pp. 421–430
173. S. Iqbal, S.A. Alharbi, Advancing automation in digital forensic investigations using machine learning forensics, in *Digital Forensic Science* (IntechOpen, 2019)
174. T. Mikolov, I. Sutskever, K. Chen, G.S. Corrado, J. Dean, Distributed representations of words and phrases and their compositionality, in *Advances in Neural Information Processing Systems* (2013), pp. 3111–3119
175. J. Pennington, R. Socher, C. Manning, Glove: global vectors for word representation, in *Proceedings of the 2014 Conference on Empirical Methods in Natural Language Processing (EMNLP)* (2014), pp. 1532–1543
176. F. Inc, fastText: library for efficient text classification and representation learning (2016)
177. M. Kusner, Y. Sun, N. Kolkin, K. Weinberger, From word embeddings to document distances, in *International Conference on Machine Learning* (2015), pp. 957–966
178. A. Vaswani et al., Attention is all you need. Adv. Neural Inf. Proces. Syst., 5998–6008 (2017)
179. P.J. Liu et al., Generating wikipedia by summarizing long sequences, in *arXiv Prepr. arXiv1801.10198* (2018)
180. N. Kitaev, D. Klein, Constituency parsing with a self-attentive encoder, in *arXiv Prepr. arXiv1805.01052* (2018)
181. J. Devlin, M.-W. Chang, K. Lee, K. Toutanova, Bert: pre-training of deep bidirectional transformers for language understanding, in *arXiv Prepr. arXiv1810.04805* (2018)
182. C. Zhang, M. Abdul-Mageed, BERT-based arabic social media authorprofiling, in *arXiv Prepr. arXiv1909.04181* (2019)
183. G. Barlas, E. Stamatatos, Cross-domain authorship attribution using pre-trained language models, in *IFIP International Conference on Artificial Intelligence Applications and Innovations* (2020), pp. 255–266
184. S. Raaijmakers, Artificial intelligence for law enforcement: challenges and opportunities. IEEE Secur. Priv. **17**(5), 74–77 (2019)
185. Á. MacDermott, T. Baker, P. Buck, F. Iqbal, Q. Shi, The Internet of Things: challenges and considerations for cybercrime investigations and digital forensics. Int. J. Digit. Crime Forensics **12**(1), 1–13 (2020)

Printed in the United States
by Baker & Taylor Publisher Services